BECOMING A FIRE- FIGHTER:

BY

BUESEME G. ABADIOFONI

CARELESSNESS CAUSES FIRE

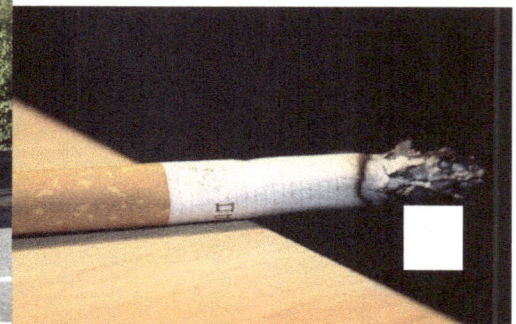

FIRST PUBLISHED BY SKYLINE PUBLISHING
(Rep. of Ireland 2016)
(T/A NEW AGE PUBLISHERS)
www.skylinebureau.com/publishing

British Library Cataloguing in Publication Data
A catalogue record for this book is available from the British Library

Copyright © 2016 by Bueseme – Godfrey Abadiofoni

ISBN : 978-0-9576498-5-9

Printed in the United Kingdom, 2016.

SKYLINE PUBLISHERS - DUBLIN
REPUBLIC OF IRELAND

SKYLINE PUBLISHERS
www.skylineBureau.com

LIST OF CONTENTS

PREFACE 8

DEDICATION 11

OBJECTIVE...11

COMMENT & RECOMMENDATION..12

CHAPER ONE

DEFINITION OF FIRE..14
UNSAFE ACTS...15
UNSAFE CONDITIONS...16
WHO IS A FIREFIGHTER & AND THE NEED FOR FIREFIGHTING....................19
WHO IS A FIRE WATCH..21
RESPONSIBILITIES & DUDIES OF A FIRE WATCH FOR CONFINED SPACES...................22
MAJOR CAUSES OF FIRE...24
FIRE CAUSED DUE TO IGNORANCE...25
FIRE CAUSED DUE TO CARELESSNESS   & COMMON MISTAKES..................25
STANDARD \MEASURES OF FIRE PREVENTION..............................27

CHAPTER TWO

THE ELEMENT OF FIRES OR FIRE TRIANGLE....................................29
FIRE WARDEN ACTIONS IN FIRE ALARM...32
FIRE FIGHTING 33
PERSONAL SAFETY IN FIRE EMERGENCY.....................................34
FIRE WARDEN INSTRUCTIONS-PLANNING BEFORE FIGHTING FIRE....................34
COMMUNICATION AND ACCESS CHALLENGES IN FIRE FIGHTING....................35
TIME FACTOR IN FIRE FIGHTING...36
FIRE SAFETY SIGNS...38
CHALLENGES OF FIRE FIGHTERS...39
WHAT TO DO IN FIRE EMERGENCY...39

CHAPTER THREE

MUSTERING AND EVACUATION POINT 41
TYPES OF EMERGENCY..42
DESIGNATED SMOKING AREA AT WORK PACE.................................43
ITEMS REQUIRED AT DESIGNATED SMOKING AREA.........................44

PHASES OF BURNING 44
HEAT TRANSFER OR WAYS FIRE SPREADS...44
FACTOR THAT MAKES FIRE TO SPREAD...46
CLASSIFICATION OF FIRE IN FIRE FIGHTING................... 47
THE ROLE OF FIRE EXTINGUISHERS IN FIRE SAFETY.............................47
METHODS OF EXTINGUISHING FIRES. 50
ACRONYM [PASS]..51
QUALITIES & RESPONSIILITIES OF A FIRE FIGHTER................................55

CHAPTER FOUR

TYPES OF FIRES AND THEIR CAUSES 57
CAUSES OF GASEOUS FIRES...58
MOSQUITO COIL FIRES AND PRECAUTIONS..60
CAUSES OF DOMESTIC FIRES WITH NATURAL FIRE WOOD....................63
CAUSES OF FIRE WITH USE OF COOKING GAS SYSTEM AT HOMES............65
DOMESTIC FIRES WITH USE OF ELECTRIC COOKING APPLIANCE AT HOMES..............68
PORTABLE POWER GENERATOR FIRES...70

CHAPTER FIVE

MAJOR CAUSES OF INDUSTRIAL FIRES..73
CAUSES OF WELDING FIRES...74
CONTROL MEASURES TO BE TAKEN DURING WELDING OPERATIONS 76
RISK OF FACILITY INSTALLATIONS FIRES 77
TYPES OF GAS DETECTORS..78
USE OF GAS DETECTOR IN CONFINED SPACE......................................76
DEFINITIONS TO NOTE WHEN DEALING WITH GAS DETECTOR- INSTRUMENTS IN CONFINED SPACE AND HAZARDOUS ATMOSPHERE..81
DO'S AND DONT'................................. 82

CHAPTER SIX

QUESTIONS TO ASK DURING CONFINED SPACE WELDING IN FIRE PREVENTION.........83
RESTRICTED AREAS IN HOT WORK TASKS...83
A SAFE DESIGNATED WELDING AREA...84
WHAT IS CONFINED SPACE?..85
EXAMPLES OF CONFINED SPACE...85

CHAPTER SEVEN

GRINDING FIRES...87
OXY-ACETYLENE CUTTING FIRES...89
ELECTRICAL APPLIANCE AND MAINTENANCE FIRES............................87
DANGERS OF ELECTRICAL FIRES...91
OIL AND GAS FACILITY MAINTENANCE FIRES..94
OIL SPILLS WILD FIRES...96
CAUSES OF OIL SPILLS...97
EFFECTS OF OIL SPILL FIRES..98
STRATEGIES AND CONTROLS FOR OIL SPILL PREVENTION...................98
CONVENTIONAL OIL SPILL CONTROL METHODS..................................98
SOCIAL IMBALANCE AS MAJOR CAUSE OF OIL AND GAS SPILLS...........99

CHAPTER EIGHT

MARINE TRANSPORTATION AND HOUSE BOAT FIRES & PREVENTIVE MEASURES.....102
OIL BUNKERS AND VANDAL FIRES..104
DOMESTIC FIRES FROM POOR WASTE MANAGEMENT................................107
TYPES OF WASTE DISPOSAL METHODS..110
HAZARDS OF POOR WASTE...110
CAUSES OF MARKET FIRES AND CONTROL MEASURES.............................111
TIMES FACTOR AND ACCESS CHALLENGES IN FIRE FIGHTING....................114
MAJOR CAUSES OF FIRE SPREAD...117
CAUSES OF SMOKING FIRES...122

CHAPTER NINE

CAUSES OF AUTOMOBILE MAINTENANCE FIRES...................................,.......126
BATTERY TERMINAL, CABLE AND WRONG CONNECTIONS FIRES.................128
CONTROLS DURING VEHICLE MAINTENANCE WORKS................................131

CHAPTER TEN

CAUSES OF OFFICE FIRES...134
HAZARDS & CAUSES OF OFFICE FIRES..135
USE OF SMOKE DETECTORS IN FIRE PREVENTION AT HOMES AND OFFICE.............136
FIRE SIGNS..136
CAUSES OF FILLING STATION FIRES...138
CAUSES OF FILLING STATION FIRES WITH USE OF MOBILE PHONES...........143
CAUSES OF SPONTANEOUS FIRES...146

CHAPTER ELEVEN

THE IMPORTANCE OF FIRST AIDER IN FIRE SAFETY..147
WHO IS A FIRST AIDER?...148
EMERGENCY RESPONSE [ER] IN FIRE FIGHTING...148
FIRST AID KIT...151
ITEMS REQUIRED IN FIRST AID KIT...152
THE NEED FOR FIRST AID BOX AT HOMES AND WORK PLACE...................153
EMERGENCY RESPONSE & RESUSCITATION PROCEDURES...........................154
CARDIO PULMONARY RESUSCITATION [CPR]...155
METHOD OF STANDARD APPLICATION..156
BASIC STEPS AND PROCEDURE TO FOLLOW...156
CRITICAL CONCERNS TO NOTE IN EMERGENCY SITUATIONS........................158
ABOUT THE AUTHOR..159
CITATION...160

PREFACE:

Discovery of fire by ancient people from time immemorial had been a good source of heat generation to man for human utilization. Fire has been used for several purposes by man all over the world. This is why fire had today become a good servant and at the same time a bad master when wrongly used. Fire had been an important source of heat energy to people, families and industries because of its enormous benefits. Fire has been used generally for domestic and industrial purposes. This include generally for heating, cooking, boiling, preservation, drying, lightening and burning of bush and domestic wastes etc, Fire has been very important to other areas such as in engineering and scientific professions, fire is used for general combustion processes e. g. used in combustion of petrol engines, air craft, engine boat, ship, cars, machines and heavy duty equipments as well as heating of other industrial systems etc.

Fire had been in use for several dynamic purposes over the years with extreme associated hazard. However it is of serous importance for people to know those common and major causes of fire, associated risk and the negative consequence that fire can impose on people, property and the environment despite its numerous benefits to people and the society when fire is wrongly used. Fire can be dangerous, destructive and deadly, several people and families have been displaced, with many been disabilities and killed with properties worth millions of dollars had been destroyed in different parts of the country from major fire outbreaks. Fire out breaks can occur in places such as Homes, Churches, Schools,

Banks, Offices, Cinema halls, Conference halls, Industries, Construction Sites, Maintenance Workshops, Vehicle and Road Crashes, Oil Facilities or locations, Filling Stations, Fuel Dumps, Fuel Tankers and Football stadiums etc. Cause of fire today has been attributed to human factors. These include people's carelessness and ignorance as well as common mistakes by people when in use of flammable or combustible substance.

Fire occurs and is most often due to how flammable materials and ignition sources are been handled, stored as well as used by people both at homes, work places and during purchase as well as transportation of petroleum and other flammable products from depots and filling stations as well as during vehicle maintenance services at workshops, road racing by drivers who carries petroleum trucks and tankers at high ways etc. So many factors have also been identified as responsible causes of fires. Today the incidents of fire disasters all over the world have greatly shown that the negative impacts or consequence of fire has been more greater than the actual benefits that it offers to the society comparing to the rates, classifications and the high level of property destructions as well as cases of death that fire have caused to people and societies globally. Numerous scenarios of fire disasters have been recorded in different parts of the world resulting in numerous lost from different types of human activities by people with no definite measure put in place to stop or reduce this menace exception of those conventionally adopted measures developed in the present day fire fighting. It is therefore the need for the populace to be adequately enlightened and educated on those common and major causes of fire as well as means of prevention both at homes and at work places in other to reduce or minimize their causes and impacts for the overall safety of humanity. To succeed in the campaign is there therefore the need worthy of the general public this include Children, Teenagers as

9

well those of our ignorant Adults especially those in schools who have not been able to acquire good understanding or ideas to the

*common and major causes of domestic and industrial fires to use this opportunity to learn and to know exactly what fire is all about, what can make fire to occur, prevention and how fire can be put off when it eventually occurs which should be the last resort in the combat against fire in actual firefighting. Today majority of the world population have come to believed that fire have become part of human existence therefore trying to put more priority attention only to how fire can be combated than preventive measures. This is one serous reason why cases of fire out break have not been adequately given attention to how it can be tackled or addressed at different levels in the society. Records have also shown from all researches and investigations that common causes of fire had been heavily attributed to people's carelessness, ignorance as well as common mistakes. This is due to how combustible materials and ignition sources are been handled, stored and used by people within the environment. These setbacks have remained a major cause of all kinds of fires all over the world. The said book on **Becoming A Fire Fighter** has been well tailored and practically designed and geared towards informing, educating and preparing people to withstand those challenges of fire, to provide good understanding about basic steps involved in fire fighting, including how to handle, store and make safe use of combustible materials, how to safely ignition sources can be handled both at homes, industries and at work place and how fire can be prevented as well as what to do in the event of any fire out break where all the necessary preventive procedures required to prevent fire have failed.*

DEDICATION

This Book is dedicated to God Almighty who inspired and strengthened me throughout this work, also to the Federal Fire Service Commission of Nigeria and Bayelsa State Fire Service Command Headquarters. .

OBJECTIVES:

The purpose of this book is aimed at providing a safe guide or platform for all class of people such as children, young adults, and adults to be aware of how fire is induced and eventually generated at homes, schools, churches, industries, construction sites, fuel storage facilities, oil and gas installations, filling stations and depots, as well as transportation and other work places through human, mechanical and environmental factors and their preventions in other to avoid destructions to property and loss to lives in our society.

11

Comment and Recommendation:
BY

MR. PAGAIBI KOKO-AGBODO

The Commandant, Bayelsa State Fire Service Command Headquarter. [State Commandant]

Every aspects of human commitments in strive towards providing sustainable measures in fight against public disasters must be encouraged by all to ensure public safety in our society at all times. This book titled, 'Becoming A Firefighter' written by the author on general causes and prevention of fire in our society is a practical step taken in a positive direction aimed at educating the masses especially young once on how best fire can be adequately prevented as well as how people can actually deal with combustible and ignition materials which have been the major causes of domestic and industrial fires. Also on how best people should be able to fight fire eventually if it occurs. It is therefore imperative to encourage the public to make good use of this book in other to invoke a positive change of attitude amongst members of the public on the use of these substances in other to reduce the high level of fire incidents and disasters as well as los to lives in our society. Causes of fire ranging from Human, Mechanical and Environmental factors have

actually occurred and resulted in various classifications of fires with resultant negative impacts caused to People, Assets and the Environment hence the need for a collective responsibility and commitments of all with strategic measures to educate and enlighten the populace upon which this book is one important step. I therefore, recommend this book **Becoming a Firefighter** *to the public for adequate use in other to provide a safe guide to people against unnecessary causes of fire disasters at homes and at work place.*

c a r e l e s s n e s s
causes Fire...

CHAPTER ONE
WHAT IS FIRE?

DEFINITION OF FIRE:

Fires may have different definitions by different persons depending on how fire is seen as well as about its behavior and characteristics but can also be defined as follows:

- Fire is defined as the interaction between a fuel, oxygen and heat at appropriate temperature, resulting in the production of flame.
- Fire is the rapid combustion of two or more combustible substances resulting in the production of heat and light.
- Fire is the rapid combustion of two or more substances which produces heat, light, smoke and carbon, these elements in their actions are sometimes referred to as physics and chemistry of fire because of the presence of the following elementary reactions and characteristics.
- It could also said to be a chemical process involving vaporization and oxidation of combustible materials,

14

- accompanied by the release of energy in the form of heat and light.
- Note: The above definitions emphasize the importance of the three basic components or elements of fire.

These include, **Fuel: Heat: Oxygen:** Before any fire can occur, the above elements **Oxygen**, **Heat** and **Fuel** must be present in an equal proportion. To achieve more clarification and understanding about how fire can occur as well as what fire is all about there is need for people to have good understanding and information as well as knowledge on how fire is usually induced and generated. The pictures bellow shows the fire triangle and how these elements combines to make fire occur as well as how fire can be put off with the removal of any of the elements when there is fire. Fire is however caused by different factors but in all ramifications it boils down to human factors. These factors have remained major causes of several infernos. These human factors include, **UNSAFE ACT** and **UNSAFE CONDITION.** Let us look in to the different definitions of the above actions and situations that when occurred can result to devastating circumstance as well as fire out breaks both at Homes, Industry, Schools, Offices, Churches, Construction Sites and work places.

WHAT IS UNSAFE ACT?

An Unsafe act; these are unsafe attitude and behavior and violation of people that can lead to injuries, property damage and death at home, industry and work sites. This acts when carried out where there is storage of combustibles and ignition source can actually cause fire. Other unsafe actions in this regard that can still induce fire include the ways at which combustible materials and ignition sources are been handled by people. These include their usage, storage and method of transportation within the environments. So many other dangerous situations can be affected by these acts but our sensitization shall be limited and anchored on those common causes of fire in particular to unsafe act because of how incidents of fire outbreak is in continual occurrence all over the world.

WHAT IS UNSAFE CONDITION?

An Unsafe Condition these are working conditions created by people, nature, with most times due to personal or management negligence. Unsafe condition actually shows a poor safety culture both at home, workplace and site, roads, homes, offices and other work places. These situations can occur due to system failures, environmental and equipment as well as machine failures, e. g inadequate or poor working environment, caring out general hot work tasks close to flammable or where flammable substances are stored or kept, e.g caring out welding in confined spaces, caring out lifting and hoisting operation within an uneven ground or unsafe environment, lifting and hoisting close to high tension power line, caring out excavation works within areas where underground facilities are laid without proper checks, caring out electrical works under rain or damped area or without adequate isolation, working under a suspended load, walking around slippery surfaces, working closed to naked cable or wires, working at height without use of harness belt etc as well as managements negligence in respect to intervention, use of defective equipments such as machine or tools, lightening of matches or lighters closed to petrol or gaseous substances, working without adequate personal protective equipment P P E, engaging incompetent and unqualified personnel at work, inadequate supervision and maintenance of plants and equipment, use of cooking gas bottles indoor, emission of poisonous gases in to the environment without adequate control measures etc are all unsafe conditions that can lead to different forms of accident to occur. The conditions at which combustible materials and ignition sources are used and kept as well as been transported are the common and major causes of both domestic and industrial fires.

It will however, interest us to note those common and major causes of fires as well as their preventions before we go in to details about their real scenarios.

So far, as we read on we shall look at those common and major causes of fire one after the other a fires both at homes and work places. Major causes of fires are actually results from human violations which is often times through ignorance and common mistakes due to people's conducts, these include how people sometimes ignorantly violates against laid down procedures and safe handling practices about combustible materials. Others causes include misuse of combustible materials in presence of ignition sources at in the homes and workplace through carelessness and common mistakes. It will interest us to remind ourselves of those common and major causes of fire and their preventive measures at the end of our discussions.

Fire Brigades Fighting Fire: **Make Call in any fire emergency**

WHO IS A FIRE FIREFIGHTER AND THE NEED?

The name a Fire Warden or Fire Fighters all refers to a team or individuals that seeks to ensure the safe condition of a place or environment where hazardous task can generate fire. Fire Fighters can also be referred to persons that are professionally trained on how best to handle fire emergency situations when the need arise. These are sets of persons whose responsibility include evacuation of people [occupants] that may be involve within a designated building or an area if eventually there is fire. One critical point to note is that both, Fire warden, Fire watch and Fire fighters are all persons vested with similar responsibility to provide fire safety for all at all times. Their services include monitoring areas of hot work activities that may generate fire. Also to ensure that fires are well monitored with the use of gas detectors as well as ensure that fires are combated and put off with the use of necessary fire fighting equipments, such as Fire Extinguishing mediums of different types, e.g Fire extinguishers, Fire Hydrant, Sand Box, Fire blanket etc readily available and functional to challenge any outbreak of fire. The task of these categories of persons can always arise when Fire The issue of fire watch is usually very specific with respect to intervention, the fire watch maintains a clear line of responsibility by showing evidence of direct monitoring for possible fire out break due to high level of uncertainty during any hot work activity within an area where there is presence of flammable substances or hazardous atmosphere and during actual fire emergency situations.

19

The Fire Warden monitors all hot work activities to avoid fire as well as directs people within a building on safe exit routes in event of any fire break to ensure safety of lives. They are also saddled with the task of fighting fire to a stop as well as rescue and evacuation of persons entrapped in fire scenarios. They also help in encouraging professional approach when responding to fire situations and use of alarm systems in real fire disasters. Fire warden are needed for various reasons. You will appreciate that it is impossible to know who could be in a building at every point in time or where they are, and therefore accounting for people is usually very difficult. There are several scenarios where little children and even grown up adult has been forgotten and entrapped within fire scenarios such as building when under a blazed due to tensions that are usually generated from emergency situations inclusive of fire outbreak. There are numerous cases where tender aged children in several occasions of fire disasters have been forgotten and allowed in to painful deaths with many others been rescued by these professional?, hence a sweep system where fire warden checks that an area or apartment is empty, if it is not safe to do so on their way out of a building. Firefighter is someone specially trained and competent enough, having basic knowledge and skills to control, and handle as well as fight all categories of fire related situations both at homes, offices, Industries, Construction sites, markets and other work places etc. A fire fighter should have good knowledge of common causes of fire and prevention as well as the need to possess the good qualities such as; He must be familiar with the operational principles of all types of firefighting gadgets such as the use of fire extinguishers, fire hydrants, have a good knowledge on the different types of fire extinguishers and   their classes of fires, have good knowledge on emergency response procedures during fire outbreak, have a good knowledge of first aid administration and evacuation procedures, to be very smart and alert, ready, dedicated and always focus, prepared and calculative [fast reasoning] efficient in communication and in coordination as the case may be during any fire situation. Many lives have been sent to their early graves due to sometimes people with over confidence and over reliance on fire fighters whose might be far away distance in real life disasters with especially those of our lazy and unmannered persons and approaches in response to real fire emergency situations therefore imperative to have well trained and responsible firefighters in the interest of public safety.

WHO IS A SAFETY WATCH?
A Safety watch is a person assigned to a job for the sole purpose of observing for hazard and monitoring work

processes to ensure safe completion of work. Only a safety watch trained in the use of Self Contained Breathing Apparatus [SCBA] shall be used for entry in to jobs where the workers in the confined space are required to wear breathing apparatus with a separate air supply and shall be ready to put on the face piece as the case may be. In addition a back up person must be immediately available to back up the safety watch. The backup person must also have breathing equipment available and know how to use it as well as to ensure that good communication is maintained between the safety watch and the person inside the of the confined space at all time during execution such task.

RESPONSIBILITIES AND DUTIES OF A SAFETY WATCH FOR CONFINED SPACE;

- A Safety watch must wear a distinctive color vest; armband, helmet, or clothing so he/she can be easily recognized as being a safety watch.

- Have two ways radio tuned to the appropriate frequency for communication with others.

- Remain immediately outside the opening to the confined space when workers are in side, maintain visual contact and communication with them at all times. Communication may be by hand signals, radio or megaphone.

- Make sure that the emergency action plan is at the site and available for review by workers.
- Ensure that no one removes blinds or shuts off ventilators or other equipment necessary for the safety of workers inside.

- Keep track of workers when they enter or leave the confined space to make certain that everyone is accounted for.

- **This includes:**

- Printing the name of each worker on the back of the daily safety entry permit.

- Verifying that workers who will be entering the confined space have signed the back of the daily safety entry permit.

- Crossing out a workers name whenever he/she leaves the confined space and will not be re-entering

-
- Ensure that SCBA is available and functional for use

- Keep unauthorized personnel from entering the confined space

- If a hot work is being done inside the confined space and the safety watch is doubting as a fire watch, he/she must have a 30-pound dry chemical powder fire extinguisher immediately at hand as well as a charged fire hose, when available.

- Be alert for change of condition and hazards that may affect the safety of workers inside the confined space

Order the immediate evacuation of the confined space whenever;

- A. a potential hazard condition arises outside the confined space
- B. workers inside the confined space are displaying the effect of over exposures to vapors of gases or extreme temperatures
- Don't enter the confined space until another worker outside has assumed safety watch duties. If entry is for rescue and the emergency is the result of an unsafe atmosphere, S C B A must be used.
- Call for assistance in case an emergency develops inside the confined space.

MAJOR CAUSES OF FIRES:

- **Ignorance**
- **Carelessness & Mistakes**

Fire is most often caused due to common and major factors.

These can occur due to the ignorance of children watching and through carelessness from smoking, leakages from gas cylinders, also by accident through vehicle collision, electrical fires, air crash, Arson which is a willful act by people this can take place through fraud, frustration and anger, and by nature e. g thunder lightening, earth quake, volcanic eruption as well as spontaneous fires.

Fire caused due to ignorance:

Fires caused due to ignorance could be said to mean all forms of fires that are caused with facts that all actions that result to the fire are usually unknown and without the understanding, knowledge or awareness of the person whose conducts and actions results to the said fire. Persons without knowledge to common causes of fires always stand a chance of causing fire outbreak than someone who has the awareness.

FIRES CAUSED DUE TO CARELESSNESS AND COMMON MISTAKES

Fires caused due to carelessness are fire usually caused with the knowledge and awareness of the persons with understanding to the possibilities of fire to occur from their conducts and actions but this is often occurred when people with such knowledge pays very little or no attention towards their conducts when executing certain task which eventually emanates in to fires. Several of these incidents could arise in situations e. g during sales of petroleum products at filling stations and especially when adequate attention is not given by filling station sales attendants while handling pump machine nozzles, smoking within filling stations, caring out welding activities within or close to filling stations, leakages from vehicle fuel tanks, refilling of petrol under running

25

engines or idling [hot exhaust, sparks from defective electrical connections, static electrical energy, smoking closed to combustible as well as electrical fires from poor electrical connections of pump machines.

GENERAL CAUSES OF FIRE:

Major and Common causes of fire has been attributed to Ignorance and Carelessness, though Causes of fire may however, depend on the nature of situation, conducts and actions which may generate fire to occur. Some of this causes include generally of the conducts and actions taken by people especially men, women and children that may induce fire to occur and this is often due to ignorance and carelessness at homes, schools, churches, hospitals, banks, markets, industries, construction sites, oil facilities, welding operations, workshops, filling stations, vessel, pipelines, during transportation of petroleum products, petrol storage tanks, foot ball stadiums, offices and work place etc in respect to how combustible materials and ignition sources are been handled and used. Common causes of fire include natural occurrence e. g. thunder lightening, Earthquake and volcanic eruption. Some of these causes are commonly induced by mistakes in violation to approved ways and procedures provided by legislation for the safe handling of flammable materials and ignition source. These include smoking in prohibited areas, flammable liquid, defective electrical equipment and spontaneous ignition. Globally there have been several incidents of fire outbreaks with multiple lost resulting from solid, liquid, gaseous and spontaneous fires with little or no attention given to the public in respect to prevention and safety awareness sensitization. For fire to occur the following elements of fire are expected to be in their equal proportions.

STANDARD MEASURE IN FIRE PREVENTION:

Summary of causes of fire and control measures:

There's plenty of air, plenty of fuel and plenty of ignition sources around construction sites, filling station, gas storage area- so we all have to be on our toes to prevent fire.

Here are some ways to keep the job from going up in smoke and flames:

- Help keep the site and work place clean and tidy at all times.
- Store combustible materials far away from ignition sources.
- Report any possible fire that you notice: open flames, sparks, and electrical equipment that appears to need repairs.
- On hot-work jobs be sure combustible are safe from ignition. Have a fire extinguisher handy for welding and cutting operations, or when open flame equipment is used.
- Help protect temporary electrical wiring from possible damage. In case of fire in or near live electric equipment, use a dry chemical extinguisher, and not water.
- Don't smoke rear flammables, No "Smoking" areas, or while refueling equipment. Make sure cigarettes and matches are out.

27

- Always use approved safety cans or the original manufacturer's container to store flammable liquids. Keep these containers closed when not in use, and never store near or passageways.
- Clean up any spills as soon as they occur. Put saturated rags into closed metal containers.
- Know where the closest fire- protection equipment is located, and how to use it. Check to see that firefighting equipment is in the clear, in proper functioning condition, and ready for instant use.

CHAPTER TWO

THE ELEMENTS OF FIRE OR FIRE TRIANGLE:

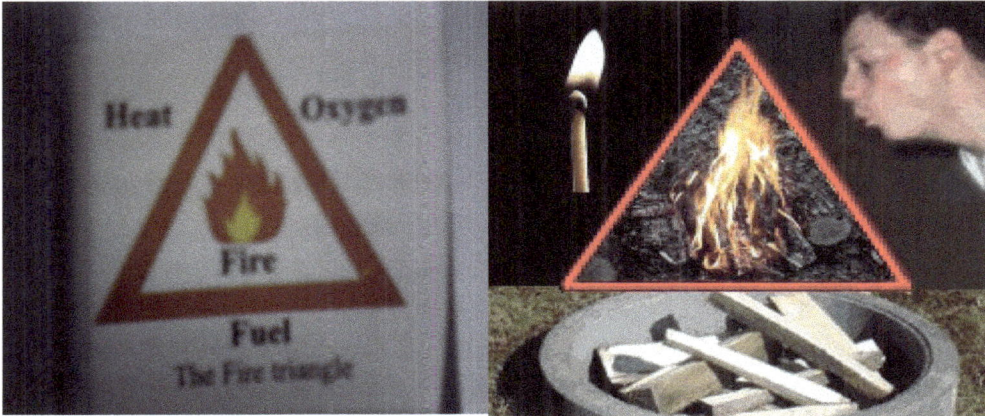

Before any fire can occur there are three elements that come together in a proportional ratio which enable fire to burn. These elements include,

Definition:

Oxygen: Oxygen is a gas which constitutes about 21% of the atmospheric air. [Fire goes off when oxygen falls below 16%]
Heat: Heat is a form of energy needed to raise the temperature of the fuel to its ignition temperature before it can burn.

[**Sources**: Friction, electrical, sparks, and radiation]
Fuel: Fuel is any combustible substance capable of burning. e .g Examples of Solid fuel are : - wood, paper, textiles, etc.

29

Liquid fuel: - Examples of Liquid fuel are petrol, kerosene,paint,or any liquefiable substances.

Gaseous Fuel: Examples of gaseous fuel are - propane, butane, acetylene, I P G, etc.

Oxygen, Heat and Fuel .

For fire to occur, these three elements must be present in their equal proportion. The pictures and demonstration above by the person shows the combination of these elements that makes fire to occur. The natural atmosphere consists of a large quantity of air with 21 percent of **oxygen** in abundant which we use in breathing. Fire cannot occur without the presence of oxygen.

Fuel could be in any form either in **solid, liquid** and **gas** e. g paper, wood, cloth, petrol, kerosene, propane, acetylene etc.

  Heat could be from any source either from naked flames, easy heat from the flame of lighted matches, candle, electric iron, heaters, stoves, sparks from two metal bodies, sparks from electrical connections, heat from the exhaust of pipe of a running vehicle engine, flame from extreme sun light [radiation] Any attempt of removing any of these elements in the presents of fire it will eliminate or put off the fire.

TOPIC SUMMARY:
It will interest us to remind ourselves at the close of this topic of those common and major actions that can easily induce fire both at homes and at work place. Basically before any fire

30

can occur definitely there shall be wrong actions and conducts by people in violation or by nature through which fire can actually occur. These include how people behave and interact with combustible materials in the presence of ignition sources within the environment e. g playing with matches or lighters in the presence of flammable substance like petrol or kerosene, caring out welding works close to combustible substance such as gas and other flammables can all cause disastrous fire out break as well as other natural causes such as thunder lightening and volcanic eruption etc.

What are the elements of fire?

Elements of fire are those substances or properties that can make fire to occur when they come together in their equal proportion or complete circle, these include Oxygen, Fuel and Heat.

MANUAL FIRE ALARM   EMERGENCY FIRE ALARM SWITCH

Fire Warden Actions in Alarm System:

- Wear your fluorescent jacket- it makes you conspicuous and easily identifiable.
- Search your area thoroughly – check store room, toilet etc.
- Report to the departmental safety officer.
- Report any problems encountered in your area [e. g people who refused to leave, people within impaired mobility who requires assistance, fire, smoke areas that could not be searched].

In the event of any fire emergency first step to take is to raise alarm and look out for the fire alarm switch or actuator and press it on to energize the alarm for signal. This is always positioned visible on walls of buildings/structures for easy identification and where people can easily gain access to them. In most cases you may see the switch covered with a glass cover. Break the glass cover to gain access to the switch. The alarm is of manual and electric or electronic types, the manually operated one is often operated with hand and by turning the handle in an anti- clock wise direction the electric is with electric current usually operated with the help of a switch button when depressed as shown above. The essence of the alarm system is to create immediate awareness of fire or other situations of life threats both at homes, offices, churches, industries and work sites.

FIGHTING FIRE:

The operations in fire fighting should include Fire Emergency Response and Evacuation Procedures. Fire emergency evacuation plan FEEP is a written document which includes the actions to be taken by all in the event of fire emergency and the arrangements for calling the fire brigade etc. As a Fire warden you are not expected to fight fire, but you are encouraged to learn about fire extinguishers. In the event of a fire, equipment is provided so that if it is discovered at a very early stage and is small, [Incipient Stage] it can be tackled in a small manner. However, in event of major fires the fire Brigade should be smartly contacted and allowed to fight such fires because of their specially acquired skills in fire fighting and their competence in fire fighting. The usual types of fire extinguisher are water, dry chemical powder and carbon dioxide. With the introduction of the current British Standard BSEN3 all fire extinguishers are now colored RED with a colored band or colored writing indicating the type of media.

Carbon dioxide C02 is **Black** - for Electrical fires

Dry Chemical Powder DCP is **Red** - for all purpose

Water = is **Red** for Paper and wood

Foam is **white** in color - for liquid/oil

Note, don't use water or foam on electrical equipment fires

When using DCP type keep yourself in fresh air because of possible fume powder that can be inhaled when adequate PPE is not worn

Personal Safety in Fire Emergency:

>Asses and be sure of been capable of fighting such fire

>Do not go in to areas containing much cloud of smoke

>Do not go in to areas where you can see established fire.

>If you find any door closed, feel it with the back of your hand before opening-if it is hot Do Not Open It, it is usually magnetic and deadly.

FIRE WARDEN INSTRUCTION
PLAN BEFORE:

> Never endanger your own life when caring out Fire Warden Duties.

> Familiarize yourself with other fire warden on your roster

> Meet with other warden and establish a plan for checking for your assigned areas. Note: wardens may agree to check their assigned are first, then convene with other warden, if warden is absent you may have check the entire area on your own.

34

COMMUNICATION & ACCESS control IN FIRE FIGHTING:

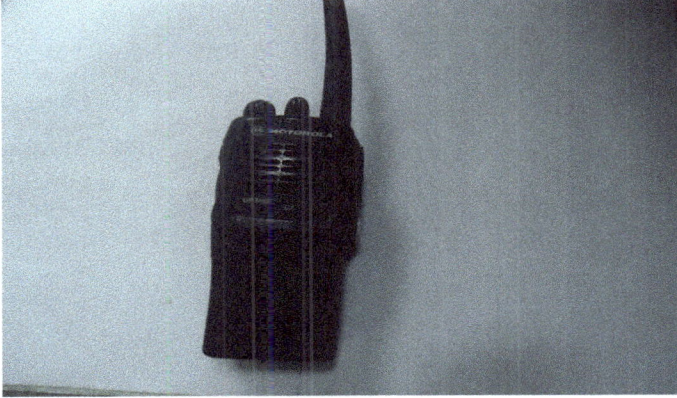

Good communication is the bedrock in any fire fighting. It matters a lot in the control of all emergency related situations this include how message or information is send from persons involved in fire scene as well as how information is been received and understood by fire fighters at the receiving end. Several lives have lost, houses and properties has been destroyed due to improper communication or inadequate dissemination of information by people during fire emergency outbreak and therefore the need for people to note, know their information are delivered, say what you mean and the response team at the other end should also understand what you mean in your communication mean should be understood during any fire out break at any point in time. Communication is power. This situation also implies to inadequate access during fire emergencies situations, many fire incident have occurred with fire fighters denied access to fire scenarios because of inadequate road net work as well as congestion of residential areas due to sometimes because of obstructions from indiscriminate congestion of houses build too close to one another in different parts of our societies. Poor planning of residential areas and business areas by government and people without proper considerations or regard to emergency situations and evacuation through provisions of adequate accesses have really caused resulted in access restrictions and made entry in to fire scene very hectic and hindered several efforts of genuine firefighters in many fire emergency situations and resulted in many lost. All buildings, residential areas, markets and public places should be build and well constructed with provisions of adequate access, these include necessary escape routes, access ways and roads so that in the event of any fire outbreak or other emergency situations, emergency response personnel as well as any ordinary other men with a good knowledge of fire fighting should also be able to gain free access in terms of rescue and evacuation in the interest of public safety.

TIME FACTOR IN FIRE FIGHTING:

The most critical fact to note in fire fighting is time, time is very important in any fire fighting situation because of the fact that every fire is usually started from small, the **incipient** stage and usually spreads or escalate to big, **inferno** stage with time and   therefore, fire should be promptly tackled whenever it is discovered and fought with time to avoid spread. Fires discovered on time, alerted and fought on time are easily put off with less stress, helps in reducing waste of resources and avoidance of spreads and lost.

The recommended principles for the use of fire extinguisher should be last resort as far as fire fighting is concern. However since causes of fire out break cannot be totally eradiated due to human activities through violations and other environmental factors hence the need for firefighting technique to be adequately introduced at all levels amongst the various classes of people, the public and especially all house hold members in away to sustain and ensure public safety at all times. People should always be alert in all tasks that they carry out so that causes of fire can be drastically reduce or avoided within our interactions with work and the environment, people must ensure that the rightful ways in which combustible materials and ignition sources are used are adequately followed and maintained to sustained and avoid unnecessary fire out break at homes and at work places. Incidents of domestic and industrial fires as well as incidents of wild fires have become unbearable circumstances that are often caused due to poor human attitude in unsafe acts and conditions which are all violations by people and have caused huge lost to lives and property worth millions of dollars hence the need for people to be timely cautious with the way we interact with combustible substance as well as whenever fire is observed. Note, whenever there is fire always raise alarm by shouting Fire! Fire!! Fire!!!

37

FIRE SAFETY SIGNS:

EMERGENCY ALARM SWITCH   MANUAL ALARM   NO SMOKING AREA

FIRE POINT   ESCAPE ROUTE   FIRE HYDRANTS HOSE

Fire safety signs are one of the vital parts of fire safety arrangement within an offices, departments and work places. Fire doors which should be kept shut are indicated with notice stating "FIRE DOOR KEEP SHUT" and smaller sign are provided upon other doors, shutters, Escape Routes, Fire Exit or Escape routes, Muster Point, Fire Alarms, Fire Extinguishers, Fire Hydrants etc, which fulfils specific functions in fire prevention and during fire outbreak. Always encourage people in your area to comply with the instructions on the signs, but if you cannot get cooperation, please call or inform your safety department for necessary intervention.

CHALLENGES THAT FIRE FIGHTERS MAY FACE:

- Fire Escape Door preventing re-entry to floor
- Locked Doors
- Difficult People
- Injured People
- Wardens absent
- Perceived risk to your self
- Not currently on roistered floor
- Forgotten your emergency Numbers

- **After checking your roistered floor [s]**
- Call your fire warden intercom phone and report on the status of your situation if you are on floors
- Take advice from Head of Fire Department
- If the Head of Fire Department is not picking his phone proceed to evacuate and advice emergency services the status of your area or floor
- Prevent re- entry in to the building until given the "all clear"

WHEN THE "ALL CLEAR" IS GIVEN

Take advice from the emergency services [e. g you might be required to advice evacuees they return to the building] make sure that you have fire warden hand book with necessary contacts and information in them at all times.

WHAT TO DO IN FIRE EMERGENCY:

First thing to do in the event of fire emergency both at homes and at work place is to ensure that alarm is raised by shouting **Fire! Fire!! Fire!!!** Stop work, on hearing alarm, turn off all

SAFETY BUESEME G. ABADIOFONI   BECOMING A FIREFIGHTER

equipment [call for immediate help] the essence of alarm is to create awareness for people around to be aware of an incident situation and preparation to move for evacuation to muster point for responsible parties to respond accordingly. The earlier fire is discovered the better it makes easier for people to fight or evacuate fire scene. Before evacuation during fire emergency first of all Look out for your personal safety, ensure that the direction you want to follow does not have obstructions or impose you to further risk or dangers. Look out for direction of wind movement or direction of fire, use appropriate nose protection devices in case of spreads of toxic fumes from fire scene such as nose mask etc.

Always remember to call the following number in event of fire emergency **122** for necessary response. While calling for help during any fire out break always make sure that your information or message is clear and well understood by third party responders to avoid confusion with regards to the basic steps taken in response to fire fighting. This include proper description of exact location of the fire scenario, e. g descriptions of short cuts to fire scene in respect to sometimes road obstructions on usually very busy roads, the types of fire and nature of fire situation as the case may be these factors are usually very important in fire fighting.

CHAPTER THREE

EVACUATION AND MUSTER POINT

Muster point is a designated area where people move from work place to gather in the event of any emergency at work place to enable emergency response team or personnel have head count of work force or personnel within a building, work sites or work environment. In event of any emergency first ensure that you know what to do and direction to follow, this is one of the first steps in emergency response procedures, on hearing of alarm, stop all you are doing, don't panic, just walk quickly to the muster point and await instructions. Though, emergencies are of different types and threats which they impose on people, always identify the type of alarm before responding there are two major types of alarm systems that are used to alert people or work force of emergency situation. These include the short intermittent

41

alarm blast and the **long uninterrupted alarm blast** which informs people or workers of possible emergencies such as Fire outbreak, Medical [Medvac], Man over board, and Release of dangerous Gases with the short alarm blast and emergency situations such as Security emergencies e. g Militancy, kidnapping or hostage taking, Arm rubbery, as well as community disturbance, this is especially with company site that are within host communities with the long alarm blast.

Below are the different types of emergencies as following:

We have Man over board emergency, this involve when someone has fallen in to water body such as river, creeks, pits, ocean etc.

We have Medical emergency, which is when someone has sustained fatal or serious accident with severe injury.

We have First Aid case, which implies to minor cases not severe e. g bruises and minor cut etc.

We have fire emergency, which implies to actual fire outbreak.

We have security emergency, which implies to all life threatening situations. e. g Militant attack such as kidnapping, Arm rubbery, Hostage taking and Community disturbance etc.

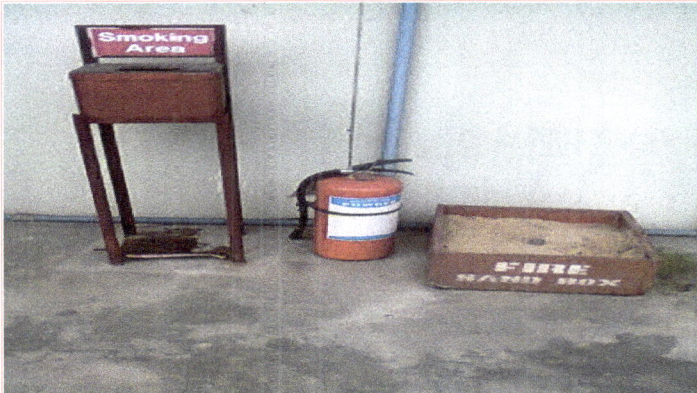

DESIGNATED SMOKING AREA AT WORK PLACE

A designated smoking area is a specific place or area designed at homes, offices, and work places especially in the industry and construction sites where chemicals, gases and other flammable substance and materials are used, handled and stored e. g Petrol, cooking and industrial gases. It is a place where people, visitors and workers are expected to stay and smoke e. g. cigarette etc, to avoid risk and danger of fire. Several fire incidents have occurred at homes, markets, workplaces especially construction sites in different parts of the globe due to how people carelessly and ignorantly carry out uncontrolled manners of smoking. This include smoking at where flammable gases and substances are stored, caring out welding, cutting and grinding operation [hot works] in gas and flammable substance area in violation to safe work procedures, poor handling of flammable or combustible materials, smoking unnecessarily at filling stations and market areas etc. It is a result of these challenges that this measure has been put in place to guide against fire hazards at homes and work place.

Basic Required items you may find at a designated area.

- **Sand box**
- **Fire extinguisher**
- **Fire hydrant/hose**
- **Smoking tray**

PHASES OF BURNING:

There is need to know the burning phases of fire when we intend to deal with fire. When fire is confined a building or room, a situation develops which requires carefully calculated and executed ventilation procedures. This type of fire can be best understood by investigation of its three progressive phases. They are:

[A] Incipient or Beginning phase

[B] Free burning phase

[C] Smoldering phase

FLASH OVER:

Flash over is a stage of fire when a room or other area become heated to the point where flame flash over the entire surface or area. This is caused by excessive heat build up from the fire itself. As the fire continues to burn all the contact the fire is gradually heated to their ignition temperature. When they reach a point, simultaneous ignition occurs and the area becomes fully involved in fire.

HEAT TRANSFER OR WAYS FIRE SPREAD:

Fire does not just occur and move from one point to the other without causes of action but this happens with the of support of other mediums which help fire to spread from one fire point to another and this is known as heat transfer. Heat transfer is like traveling birds that takes seeds from one point of location to another. Several cases of fire escalation have occurred with destructions done to properties that do not even have immediate contact with where fire started. Many houses and properties have been burnt to ashes because of closed proximities between houses and properties through the help of heat transfer. The following means are some common forms of heat transfer during fire outbreak.

They include the followings:

- **Conduction**
- **Radiation**
- **Convention**
- **Direct flame contact**

The existence of heat within a substance is caused by molecular actions. Thus as the vibration of the molecules become more intense the heat becomes more intense too. Since heat is disordered energy, it never remains constant but continually transferred to object of higher temperature. The colder of the two bodies in contact will absorb heat until both objects are the same temperature.

CONDUCTION:

Heat may be conducted from one body to another by direct contact of the two bodies or by an intervening heat conduction medium. The amount of heat that will be transferred and its rate of travel by this method depend upon the conductivity of the material through which is passed.

RADIATION:

This method of heat transmission is known as radiation of heat waves. Heat and wave are similar in nature but they differ in length. Heat waves are longer than light waves and they are sometimes called infrared rays. Radiated will travel through space until it reaches an opaque object. As the object is exposed to heat it will in return radiate heat.

CONVECTION :

Convection is the transfer of heat by the movement of air or liquid. This movement is different from the molecular motion discussed in conduction. When liquid or gas is heated, they begin to move within themselves. For example-when water is heated in a glass container, an upward movement within the vessel can be observed through the glass [the addition of saw dust to the water will make this movement more apparent]. As the water is heated, it expands, grows lighter, and move upward. As the heated air moves upward, the cooler air takes its place at lower levels.

DIRECT FLAME CONTACT

Fire also spread through a material that will burn by direct flame contact when a substance or material is heated to a point where flammable vapor are given off. These vapors may be ignited. Any other flammable material, which is in contact with the burning vapor, may be heated to a temperature where it will ignite and burn.

FACTORS THAT CAN MAKE FIRE TO SPREAD:

Fire spreads easily by conduction, convection and radiation within a building due to the following:

- >Delay in discovering the fire
- >Large amount of combustible materials
- >Lack of fire resisting structures
- >Opening in the floors and walls

- >Rapid burning of dust deposits
- >Flowing oil, fats and hydrocarbons
- >Combustible finishing and fabrics
- >Combustible construction of building

CLASSIFICATION OF FIRES:

To be competent ard successful in fire fighting it will require people to have good and adequate knowledge of the various types of fires, their classifications and extinguishing mediums as well as methods.

Types of fires include the following;

Solid Fire- A   e. g paper, cloth, wood, solid material pieces etc.
Liquid Fire- B   e. g kerosene, petrol, diesel, cooking oil etc
Gaseous Fire - C   e. g Methane, LPG Gas, propane gas, acetylene etc
Metal fires- D - e. g fires involving zinc, lead, copper, aluminum
Electrical Fire – fire involving electrical cable, conductor and insulator.

THE ROLE OF FIRE EXTINGUISHERS IN FIRE SAFETY:

Firefighting begins with the introduction of basic fire fighting skills with the use available resources and firefighting equipments in line with operational principles and policies as well as control measures and procedures that are conventionally adopted in use globally for the prevention and protection of lives and during any fire out break at homes and work place. The essence of providing fire extinguishers at homes and workplace is one fundamental safety measures put in place in view to protection of lives and property against fires outbreak at homes and workplace. The use of fire extinguishers at work place and homes is a legal requirement that has been well practiced around the globe. This requirement has been implemented to enhance sustainability of safety, to enhance protection of life and property at homes and work environment. The importance in use of this equipment cannot be emphasize upon which today it has become legal requirement as part of construction requirement. The major work of this medium is to enable people fight fires whenever there is fire outbreak especially when fire is observed early enough before it spreads from incipient stage (small fires) in to inferno stage [big fire]. These equipments in their actions have helped in putting off several industrial, homes and work place fire situations across the globe. It is however a requirement by legislation for every house hold, offices, complex, stadiums, churches, workshops, construction sites and business outfits to have these equipments in place for purpose of safety of lives and property in the event of fire. Causes of fire are numerous and depend on the nature of material that may constitute the actual burning of the fire. It is also of note that majority of people including a high percentage of the masses still does not understand how to use or operate the fire extinguishers and this has lead to several lost to lives and property destructions from even little fires to big fires that could have been put off with these equipments. It is therefore the need for the masses to be properly informed and educated on how to use fire extinguishers in the event of fire so that incident of fires can be adequately tackled when they occur. So many types of fire extinguishers are also in place for use with their specific applications depending according to the type of fire that may occur. Common types of fire extinguishers in present use includes, Carbon dioxide [C02], Foam type, and Dry Chemical Powder [D C P] used for both electrical, metals, solid and liquid fires, Foam type for gaseous fires and water which is commonly used in fighting of solid fires. Something very important to note, is time which is the highest factor in fire fighting. Whenever fire is observed on time and tackled on time definitely such fire can be extinguished on time while delayed attempt will result to spread from one point to another either through conduction, convection and or radiation of generated heat within the fire scene. Also to note, before embarking on any fire fighting, first of all assess the fire scene to be sure of your personal safety, if you cannot fight the fire then you raise alarm and better keep off the fire scene and call for fire service department for response. Always ensure that you are with fire fighters telephone numbers or with the police number. In the event of any fire call 911 for help and you should be able to provide the fire fighters with the exact information about the fire including the type of fire e. g solid, liquid and gaseous fires. This will enable fire fighters know the exact type of extinguishing medium to be used on such fire.

METHODS OF EXTINGUISHING FIRE:

Fires are usually put off with the basic firefighting principles, procedures, methods as well as techniques. The extinction is the principle of eliminating one of the elements of fire from the circle or triangle of combustion. Under the theory of fire triangle, there are three methods of fire extinction, which include **starvation, Smothering and Cooling.**

- This is the removal of either of the three elements in fire triangle e.g removing the Heat source will reduce the temperature of the heat generating the continual burning of the fire, this is by the use of water and is called **COOLING.**
- Removing of Oxygen [cutting off air] reduces oxygen source which helps to support the fire to continue, this is called **SMOTHERING**
- And removing the source of burning material from the fire [avoiding it from further spread e .g petrol or gas] this is called **STARVATION**.

THE OPERATIONAL PRINCIPLES OF FIRE EXTINGUISHER WITH THE ACRONYM [PASS]:

Fire extinguisher is a legal equipment, required by law to be part of house hold features/properties and is meant for fighting little fires when they unexpected occur. One important point to note is fact that these equipments cannot just function on their own without the intervention of people to operate them. This also involves how vigilant people can quickly observe fire outbreak including how prompt people response to fire situation as well as how this equipments is used. This intervention plays a major role and determines how fire can be effectively tackled and put off when they occur. This process however becomes very important because of the reason that time has been considered the most important factor in fire fighting. Whenever there is a delay in discovering fire, or delay in responding to fire the said fire would have

51

every chance to escalate from their little stage to inferno therefore the need for people to be vigilant, alert and be fully acquainted with the use of fire fighting systems so that fire can be actively tackled and put off in any fire emergency. Another important worries is the fact that almost 70% percent of the people does not actually understand how to operate this equipment and also about 90% percent of the majority of people having these equipments in their houses does not understand the skills, lacks the knowledge to operate fire extinguishers in the event of actual fire outbreak. Though, this is sometimes attributed to fear and tension when people come in actual contact with fire. Cases of several fires have occurred at different places with difficulties of people including house hold heads in the operation of these equipments and this has led to loss of lives, houses, markets, filling stations, fuel storage facilities, premises, workshops, industries and workplaces been burnt down, hence our concern to create awareness through enlightenment and sensitization campaigns to educate the public on how best people can effectively operate theses equipment for the purpose of public safety. Fire does not actually give warning or notice when to occur therefore the need to ensure that there is always public preparedness against fire whenever it occur. Another funny facts is that majority of people, including government officials only provides these equipments in their offices, homes and workplaces only to enable them meet legislative requirement to avoid been penalized or embarrassed of offence in case there is fire. In other to operate the fire extinguisher, there are some features on the equipment and applicable principles associated in its operation. These include the method and the operational principles used and when followed accordingly fire can be put off and when such fires are discovered on time and tackled instantly without delay. The operational principle is referred to the acronym **PASS** which shows the various steps and actions required to be taken with the use of fire extinguisher when there is actual fire.

USE OF THE ACRONYM [PASS] IN FIRE EXTINGUISHER OPERATIONAL PRINCIPLE STANDS FOR THE FOLLOWING:

The following alphabets PASS represents the action words in the use of fire extinguisher during fire emergency.

The **P** in the acronyms stands for **Pull** the pin at the top. The pin on the cylinder top helps as manual device that is used to lock the lever from accidental activation of the valve to prevent any unauthorized release of the dried carbon powder or other fire firefighting substances inside the cylinder. The pin is usually pulled out to allow control of the lever in other to enable the valve open to release the powder substance through the hose to extinguish fire.

The **A** in the acronyms stands for **Aim,** always aim at the base of the fire. When the hose is handled during fire fighting, it should be aimed or directed to the base of the fire in other to create effective disposition of the powder substance to the base. The compressed cylinder pressure should be directed towards the base of the fire and the fire will go off.

The acronym **1st S** stands for **Squeeze** the lever, the lever under the handle should be squeezed upward after the pin has been removed or pulled out while the hose is aimed at the base of the fire. Squeezing the lever helps to open the valve for the release of the powder substance for actions.

The **2nd S** in the acronym stands for Sweep the base of the fire with the hose side by side; this refers to how the hose should be handled and used. The hose should be positioned at the base of the fire and sweep side by side, this will enable the pressurized powder substance to spread to every part of the fire at the base to off the fire. Always ensure that fire is properly put off before leaving any fire scene to avoid re-ignition.

KNOWING THE ELEMENT OF FIRE SYSTEM:

To achieve a successful fire fighting at all times fire warden should be aware of the following system where appropriate to his location:

1. *Changes in floor layout of buildings*
2. *Corridors marking and signs*
3. *Fire doors [including those with self closing devices, release devices and smoke seals]*
4. *Storage arrangement and house keeping*
5. *Door furniture and security devices*
6. *Floor resisting walls, floor ceilings*
7. *Fire safety signs*
8. *External staircase*
9. *Emergency lighting*
10. *Wall coverings and Floor Covering*

QUALITIES AND RESPONSIBILITIES OF A FIRE FIGHTER:

Fire Warden/ Watch/ Fighters especially in the industries and construction sites shall be always alert for: Changing conditions in hot work or on deck above or below the area that increases the fire hazards and report these changes to the PIC Person In charge or shut down hot work if warranted. Watching for falling hot slag and spark object on areas below and around the hot work.

Operating the fire extinguisher and other available firefighting equipment. Inadequate fire extinguisher and fire pump running hose.

Fire fighters are also expected to possess some levels of first aid and medical qualities due to the nature of life threats that is sometime encountered during fire emergenc+ situations. The absence of experienced fire fighters with medical good knowledge have led to the death of several victims that could have been saved if necessary attention and recommended rescue skills and procedures would have been applied at time of incidents. These are the reasons why a qualified fire fighter should possess good knowledge of first aid and medical experience to provide a safe platform for the overall safety of people in their profession.

HOW MANY TYPES OF FIRES DO WE HAVE?

Types of fires has been very funny to good numbers of people in the public where most times people tend to wonder how truly can this be for different types of fire to exist, this to majority of the populace does not actually make any serious

55

importance to them at the same time good numbers of these categories of people only sees fire as fire and nothing more. But the truth remains the fact that at any point that there is fire definitely there should be a cause for such fire to occur as well as the types of material substance that can actually constitute such fire in to burning this is where different types of fires comes to play. Let us look forward to have clear view of what the different types of fires may be.

CHAPTER FOUR

TYPES OF FIRES AND THEIR CAUSES:

Classifications of fire and their causes have been identified. These include **SOLID**, **LIQUID** & **GASEOUS** fires and their causes which have been attributed to human, natural and environmental factors. These causes are often due violations by people from unsafe handling procedures of combustible materials and the way they been used. These causes are usually very common especially during transportation of gas cylinder bottles by road. One major factor that increases chances of road transport gas fire is often challenges of poor or bad road conditions as well as when compressed gas bottles are been exposure to direct heat sources such as extreme heat radiated from sun light and other direct sources of heat energies which should be given necessary attention during their transit at all times.

CAUSES OF GASEOUS FIRE:

The causes of gaseous fires are very many but with major hazards involved during gas cylinder transportation. These hazards can actually occur when gas cylinder bottles are poorly or wrongly been carried and especially when they are carried and kept unchained to their rags or caged during transit. Another critical situation to consider during gas transportation is when gas bottles are newly refilled at depots and allowed to move freely during their transportation, this is in regard to how random and volatile is gas under pressure. Major hazards involve in gas transportation include fire and violent explosion hazards from pressurized cylinders when they are not given adequate attention during transit. This is often caused due to violations by users and transporters to safe handling procedures of gas systems, these include, poor usage, inadequate care during transportation as well as unsafe storages. Several lives have been lost, property destroyed with incidents of disabilities and displacement to families in use of gas energies because of inadequate use resulting in massive destructions to lives and properties from gaseous fires and explosions which have occurred in different parts of the world hence the need for drivers and transporters as well as users of gas energy systems to be extremely careful when in use and handling of gas energy systems.

CONTROL MEASURES:

- The following controls should be noted during handling, use and storage of gas cylinder bottles at home and work place;
- Ensure that gas cylinder bottles are kept out of heat source
- Ensure that gas cylinder bottles are kept in upright position
- Ensure that gas cylinder bottles are installed with flash back arrestors, hoses and gauges are in good conditions
- Ensure that gas cylinder bottles and hoses are constantly monitored for crack or leakage
- Ensure that before you put on the gas for use matches or lighter in closed and ready with you to avoid delay, this has caused several domestic fires.
- Ensure that gas cylinder bottle is fitted with protective cap cover to prevent unauthorized use or against children
- Ensure that only person that knows how to use gas system should be allowed to use it.
- Ensure that newly filled gas bottles are properly transported with trolley and use gas only when it has settled down after transportation from gas depot.
- Store gas cylinder bottles at cool and well ventilated area with a shade.
- Separate gases and licuid as well as oxygen and acetylene.
- Store flammable liquids, paints and thinners in proper containers.
- Protect area around hct work with fire blankets
- Ensure that vehicles and trucks used in transportation of gas cylinders are in their safe conditions.

59

CAUSES OF FIRE WITH USE OF MOSQUITO COILS, CANDLE AND BURNING OF DOMESTIC WASTE AT HOMES:

First of all let us look at what domestic fire is all about. Domestic fires can be generally referred to as general house hold fires. What are those causes of domestic fires? Many factors have been identified and responsible to the causes of domestic fires. Domestic fires have raised several categories of buildings and premises down and caused untold hardship and havoc to millions of families around the globe due to most time common mistakes and carelessness.

THESE CAUSES INCLUDE CARELESSNESS, IGNORANT AND COMMON MISTAKES.

The way and manners at which individual and family members do engaged themselves in the use and handling of flammable materials within ignition sources at homes is a major factor, these include how fire wood [Solids materials] kerosene [liquid] are handled during cooking, lightening and heating purposes, worse of it is the case of adulterated kerosene that are in common use at homes. The uses of adulterated kerosene at the village setting by majority of villagers who could not actually afford use of gas energy is actually due to cost implications which have also led many families and members in several untold hardship as well as shown lots how difficult fire situations are extremely difficult to handle when they occur. Other common causes at the local setting include burning of unwanted domestic wastes around homes, e. g burning of grasses and other unwanted waste materials, condemned vehicle tyres which is sometimes used for lightening purposes and fun making especially during festive periods such as Christmas and New Year celebrations at homes. Domestic fire is commonly induced when these unwanted materials are burnt with their hot flames in close proximity to houses or residential areas with their heat easily transferred with the aid heat transfer. How this occurs is that the heated or burning particles of the waste from a fire scenario is usually carried and transferred to other surrounding with the help of wind movement and in many case through direct heat radiation. This is however, similar to how birds carries seeds from one part of the environment to the other. Another major factor that causes domestic fire is also the poor use of portable power generating sets at homes, this occurs when petroleum products are poorly handled or used either during refueling of power generating sets and cooking stoves. This is however, very hazardous especially when generators are kept running, [hazards of hot exhaust pipe] Records from several fire incidents have shown that numerous cases of domestic fires are caused by people through carelessness, ignorance and most times common mistakes therefore need for the people to be fully enlightened and to be extremely careful when dealing with these hazardous materials at homes and within the environment. Causes of major wild fires across the globe are sometimes fires generated due to escalations from burning of minor domestic waste.

Control Measures:

1. House hold members should be adequately educated on proper ways to handle and use flammable materials at homes especially where ignition sources are kept closed to them.

2. Sending of children in the purchase of petroleum products at homes should be avoided

3. Flammable materials should be stored far away from where ignition sources are kept.

4. Portable power generators should be switched off during refueling

5. Filled petrol containers should be properly tightened and kept far away from the reach of ignition sources or children.

6. Both ignition sources and combustible materials should be prevented from the reach of children.

7. Combustible materials should be stored far away from heat sources such as direct sun light and naked flames.

8. Oil and petrol spills should be properly cleaned to avoid fire

9. Whenever fire is discovered at home if it is electrical fire make sure you immediately switch off the main switch

or remove the fuse from the fuse box to separate current supply from source

10. Do not temper to handle any electrical appliance with damped or wet hands this will increase chance of electrocution.

11. If anyone is been electrocuted avoid live stick or metal items in attempt to move the victim, always use an insulated material or dry wood in such case to avoid electrocution.

CAUSES OF DOMESTIC FIRES WITH USE OF NATURAL FIRE WOOD:

Fire wood as a natural source of heat energy has been in used over decades and is used for general heating purposes such as domestic cooking, heating and burning of bush etc. Fire wood has been in use especially at the local setting from time immemorial till present time. This source of energy has been very helpful and affordably abundant in our environment yet, with associated potential hazard that is most times ignored by people and this have remained one common cause of several house hold fires. Several house hold fires have resulted and caused havoc to many lives especially those at the village setting where the use of fire wood is very common and the fact that majority of the people leaving could not afford the use of gas energy, people leaving bellow standard especially those leaving in abject poverty. The way and manners that people use and store fire wood at homes by people have severally resulted in great loss.

CAUSES: Some of the causes has been traced due to poor use of fire at homes, storage, storing of fire wood too closed to cooking areas, not been careful when cooking with fire wood, extreme heat generated from fire wood have caused several ignitions when heat sources are in close or direct contact with other combustibles due to extreme heat radiation, people's ignorance on the hazards of heat transfer, as well the dangers of placement of combustible materials too close to lighted fire wood, drying of cloths over lighted fire wood by hanging wet cloths over high flames this is a very common practiced at the village setting especially during raining season where people usually finds it extremely difficult in drying cloths, including how fires have been ignorantly generated due to heat transfer from glowing embers and ashes of natural fire woods during actual burning with the help of breeze hence the need for appropriate caution for all house hold members to be extremely alert during use of fire wood at homes to avoid domestic fires for overall safety of lives and property. Note, all flammable materials should be kept and stored far away from ignited fire woods to avoid fire spread and spontaneous escalation of fires at homes.

CAUSES OF FIRES WITH USE OF COOKING GAS SYSTEMS AT HOMES:

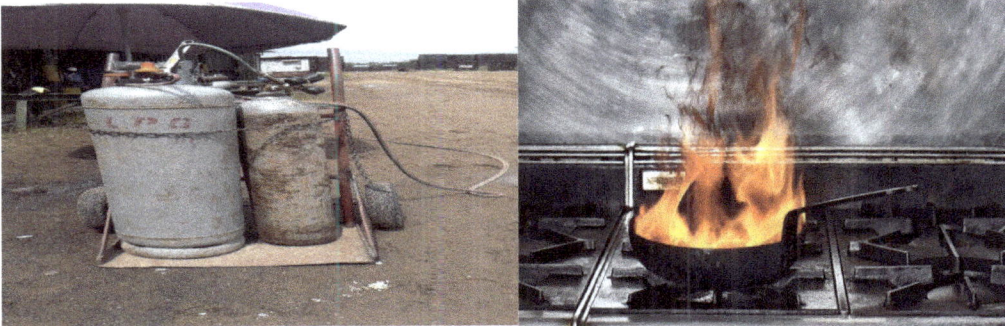

The conventional use of gas energy in domestic cooking have become imperative due the high level of advantages and energy efficiency over other sources of heat energies that are presently in contemporary used for cooking at homes. Though, gas energies are generally used for heating and other domestic and industrial purposes with enormous quality and importance. Gas energy is environment friendly and does not impose health challenge to people and the environment but is however, associated with extreme potential hazards while in their use. These problems can be caused due to poor handling by human and system failures, these include gas leaks either from, defective or loosed hose, over filled gas cylinder bottles, defective control valve, bad cylinder bottle, poor storage may be wrong positioning, [not positioning gas bottle up right], exposure of gas bottles to hot surfaces e.g heat from extreme heat from sun radiation, transportation hazards, poor lightening during use [delay in lightening], poor timing, keeping gas cylinder valve opened for too long before lighting, lightening with a faulty matches/lighters], use of un stabilized pressurized gas cylinders which have caused several gas explosions and fires. This happens when the random natures of compressed gas pressure is disturbed or shaken which eventually gives rises to expansion in gaseous state. This increase in temperature rise by the gas in the cylinder bottle results in destructive and violent explosion when the bottle integrity could no longer hold the enclosed pressure.

Note, that hazards of gas explosion and fires can never be compromised, gas cylinder explosion is usually like rocket speed and deadly when they occur therefore the need for people to be more careful in their use despite the different types of safety features that has been newly incorporated in modern gas systems for prevention of unnecessary gas fire emergencies.

WARNING AND CONTROL MEASURES:
- Make sure you buy and use gas systems with automatic self ignition lighters to avoid issues of delay in lightening which have remained major cause of several gas fires at homes.
- Cooking gas cylinder bottles should be very strong and reliable in their make and product. Let it be of quality types.
- It should be installed with flash arrestors.
- It should be accompanied with protective cap.

Let it carry functional regulator and gauge meter indicator
- Hoses should be thoroughly checked for crack before and after use.

- Cylinder bottle should be checked for crack and leakage at all times.
- Gas leak detector should be installed at homes to monitor any gas leak at homes to avoid unexpected fire and explosions.

-
- Gas cylinder bottles should not be over filled above required gauge level.
- Gas cylinders bottles should not be exposed to direct heat source or radiation of the sun light.

-
- Gas cylinder bottles should be transported with rag/trolley and kept vertically up right during storage.

-
- At the end of every use cylinder valves should be properly looked and kept covered under the cap to avoid unauthorized use.

-
- When to put on cooking accessories such as cooking gas cylinder make sure that the valve is opened slowly to avoid explosion

-
- Make sure your lighter or matches is closed to you before opening the control valve to avoid delay during lightening for this have caused many explosion and fires at homes.

- Fire-wood ashes should be properly kept after cook within cooking environment to avoid spread either through wind movement.

CAUSES OF FIRES WITH USE OF ELECTRIC COOKING APPLIANCE AT HOMES:

Electrically operated systems as well as cooking appliance are of different types and different make and products, these appliances are in common use today, because of their advantages in respect to economical factors as compared to other sources of heat generating appliance. These appliances become very important at homes because of the fact that they are electrically operated with availability of electric power in abundance everywhere to operate them with view to fewer expenses in their use. However, apart from the enormous benefits with use of thee appliances yet there are still lots of potential hazards involved in their use which have resulted in several domestic electrical fires. Some of these appliance include electric cooker or stoves, heaters, ring boilers, electric kettle, water dispenser, Television set, electric ovens, washing machines, electric grinders etc, with their associated hazards such as electrocution [shock] which usually occurs as a result of defective electrical appliance, defective electrical connections, short circuits, naked cables, using appliances under wet ground or surfaces, inadequate earthen of electrical appliance which most times attracts thunder lightening etc, these challenges have all become a major source of worries amongst other causes. Other causes include how electronics and other electrical appliances such as electric pressing iron are used with most times you see people leaving already powered pressing irons unattended to while on cloths or combustible materials, others include poor connections and use of electric washing machines, this is sometimes the way we see them been connected and carelessly used by people at homes especially when adequate attention is not given the way it ought to according to manufacturers recommendations. For instance is the use of high voltage appliances against recommended house hold wall sockets with regard to power rating on these appliances at homes, good number of domestic fires in regard to this practices is that when high voltage appliance is loaded on to low voltage cables and wall sockets the over loaded voltage generate an internal heat inside the socket cables and subsequently result in melting of the insulations, melting of the insulated cables result in bridge between the negative and the positive wires which eventually causes sparks and fire especially when exposed to combustible substance. Other factors include use of unqualified or incompetent electricians in the installation and wiring works at homes, poor interface between NEPA power sources and portable generator systems, use of defective change over control switch, using electronic appliance and equipments under thunder lightening, use of defective electrical appliances, poor insulation of electrical cables and wiring at homes, use of substandard electrical fittings with regards to power rating due to cost implications, over loading by applying too many appliances on to a particular wall socket as the case may be without consideration to their implications. This has remained a major cause of several house hold electrical fires and therefore people should be cautious with the use of electrical appliances at homes.

CAUSES OF MAJOR ELECTRICAL FIRE:

PORTABLE POWER GENERATOR FIRES:

Uses of portable power generators especially the locally named [one man carrier] as shown above used at homes, offices and work places has become very important with numerous help to people because of their mobility. Power generators despite its advantages yet are associated with great potential hazard in their use due to the nature of how they are sometimes used at homes and other places by people. The uses of this equipment have increased due to incessant power failures different parts of the world. Inadequate power supply situations have lured many house hold into the compulsory use of these portable generators both at homes and work places. One important fact to note with the use of this system by people is the fact that majority of people has been threatened with the fear of theft of this power systems in their homes. Because of this fear many families have been ignorantly engaged in the use of generators indoors both at homes and workplace without knowing the effects and harms it has on people and the environment. Several incidents of inhalation of carbon monoxide poisoning generated from these generators and consumed by families including fire outbreaks have occurred at many homes because of people not having awareness of these hazards. Lots of houses and offices have been set ablaze with majority of these houses raised down due to using generators indoors. The causes of these fires actually depend on how people use and keep these systems at homes, other causes include wrong and poor electrical connections, use of defective appliance, over loading, sparks from generator battery cables and terminals especially those using battery and starter motors, defective socket plug on generators, use of low capacity cables and wires, inadequate installations, short circuits, heat generated from hot exhaust of generators which when in contact with flammable substance can generate fire, sparks from loosed plug high tension cable in presence of fuel leak, as well as naked flames during refueling of generators etc.

CONTROL MEASURES:

- Prior to the use of power generator users should ensure that generators are kept out of leaving rooms [outdoor]
- Ensure that there is no fuel leakage around the generator before putting it on.
- Ensue that generator is kept well ventilated area
- Make sure that flammable substances are not placed or positioned close to generator area especially when the generator is on or running
- Ensure that cable cords are of recommended capacities
- Make sure that sockets of generators are of same size with the external cable plug in other to avoid sparks from loosed contact between sockets which have caused several fires from sparks.
- Make sure that junction box where electrical connections are made should have a very good and firm joint contact.
- Fuse box and cut out fuses should always be of rated capacities and recommended qualities.
- Be always vigilant of fires from glowing fuse contacts points.
- Battery cables and terminals should be properly tightened and secured with rubber insulation cover to avoid contact with earth body of generators.
- Always ensure that plug caps and high tension cables are always secured and properly insulated
- When refueling make sure that generator is shut down
- Ensure that generator exhaust is always directed far away from house.

CHAPTER FIVE

MAJOR CAUSES OF FIRE WITH INDUSTRIAL ACTIVITIES:

Cause of Industrial Fires:
These include generally of all electrical and hot work activities which is often carried out in the industries, welding and fabrication workshops and construction sites etc.
FIVE BASIC STEPS for FIRE PREVENTION IN WELDING/GRINDING AND CUTTING OPERATION.
These include adequate use of caution sign post, permit to work, fire extinguisher, ensured gas test & monitoring, containment of sparks, and knowing the emergency phone numbers of emergency response team.

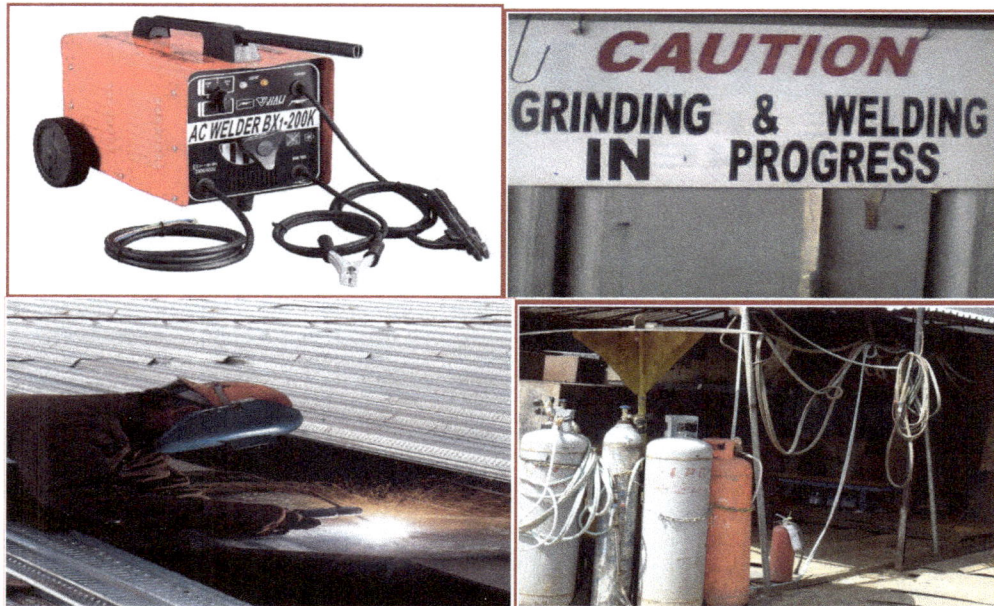

CAUSES OF WELDING FIRES:

Generally, welding is a process of joining two or more metal pieces together with the use of welding machine and filler electrode, it is usually carried out by specially trained personnel called Welders. Welding is of different types with majority of the types usually found at construction and fabrication sites. In welding operations there are so many associated hazards and risk involved, these hazards includes the followings;

Fire

Explosion

Electric shock from electric welding machines

Emission of toxic fumes from heated metals

Heat stroke

Arc eye

Extreme sweat from heat radiation which most times result in mild electrocution of the human body through conduction when electric welding machine is not properly earthed as the case may be.

CAUSES:

The above dangers can occur and lead to fire and fatal death at work place especially when safe recommended welding procedures is violated by people and most especially where fire protection devices are not in place during task execution at work place. Welding activity is a dangerous task. Some of its problems can arise when generally work procedures is violated, this include caring out welding works without obtaining permit, the way material and equipments are wrongly or poorly used, not following safe use procedures, failure to check isolation before welding especially in confined space and without gas testing, as well as exposure of cylinder bottles to extreme sun radiation, placement of flammable materials too closed to welding area, use of incompetent personnel in task execution, welding without effective supervision by competent supervisors, welding on pipe lines with adequate precautions, not putting in place fire fighting mediums, e. g fire extinguishers, etc, placement of combustible substance closed to where welding works is been carried out etc.

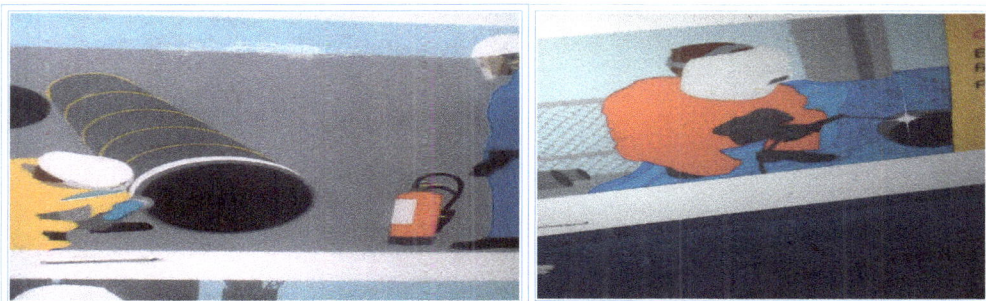

USE OF FIRE EXTINGUISHER:

75

CONTAIN SPARKS WITH FIRE PROOF MATERIALS
Control measures to be taken during welding operations:

These include:

- Before the commencement of any welding work, there should be adequate risk assessment to identify potential risk areas on the work environment, inspections to be made on storage systems of flammable substance and materials, facilities, flow lines e. g oil and gas pipe lines, open drains, vessels, and

- channels should be inspected and where necessary flushed or purged with steam/water etc and capped off.
- Wet fire resistant blankets shall then be positioned to surround the work object and isolate it from all adjacent equipment.
- Flammable substances should not be placed near welding area.
- Adequate ventilation including barricade should be provided within welding area
- Fire extinguisher shall be provided at welding area

RISK OF FACILITY/INSTALLATION FIRES:

A fire watch shall be posted outside the enclosure and make sure that all sparks and slag are contained within the blanketed area. He shall also ensure that the blankets are kept wet for the duration of the work. Prevention of Facility fire is the sole responsibility of facility owners, Asset controllers and security operatives as well as adequate monitoring of gas facility with the use of fixed or permanent Gas detectors to maintain and monitor possible gas leak with regard to atmospheric changes at certain environment. Fire is usually caused by people due to human interactions with the natural environment, natural occurrence and violations through carelessness, ignorance and common mistakes where adequate care should be provided at all point in time especially within oil and gas installations where potential hazards of fire is always present and high.

12 LIFE SAVING RULES NUMBER TWO:

Conduct gas test when required.

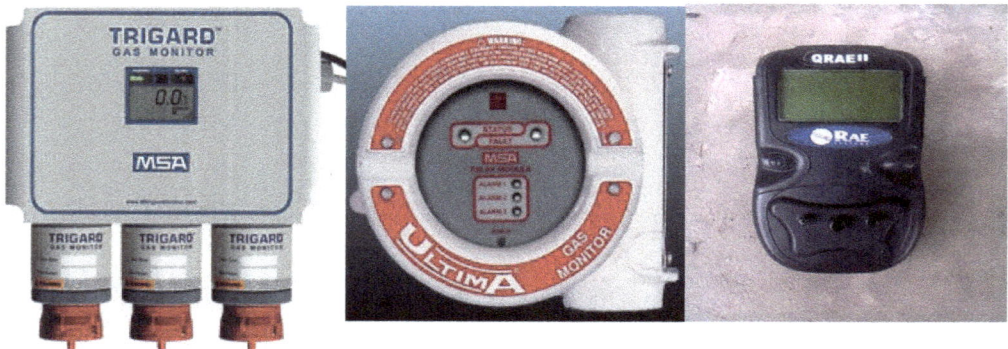

FIXED & MOBILE TYPES OF GAS DETECTORS USED IN MONITORING OF GAS LEAK, HAZARDOUS ATMOSPHERE & CONFINED SPACE WELDING OPERATIONS:

THE USE OF A GAS DETECTORS IN CONFINED SPACE TASK & HAZARDOUS ENVIRONMENT:

OBTAIN VALID PERMIT BEFORE CONDUCTING A GAS TEST
HOW TO CONDUCT A GAS TEST WITH GAS DETECTORS:

The use of gas detector in fire prevention when working in a confined space and hazardous atmosphere;

INTRODUCTION TO USE OF GAS DETECTORS:

The use of gas detectors has been in existence over the years and comes to play due to the high level of fire and explosion hazards in the Oil and Gas exploration and other industrially operated activities. The presence of different types of dangerous gases in the atmosphere and their changes to atmospheric conditions especially where oil and gas task is carried out with the possibility of asphyxiation as well as fire and explosion hazards to personnel, assets and environment are the reasons why the use of gas detection equipment have become necessarily important for the safety of Personnel, Assets and the Environment at work place. Incidents of fire out breaks and disasters ranging from domestic and industrial fires from gas related problems with lost to lives and properties worth millions of dollars have occurred at different intervals in different parts of the continents and most especially within industrially operated regions with these disasters most times caused due to failures in parts of work personnel not been proactive and non compliance to safe work procedures as well as not been adequate in the detection of gas presence at the systems and the environment they work. This has been so far very common in the exploitation and production task of the natural OIL and GAS resources in different parts of the environment. Some of the major causes of these hazards include Gas seals, Flanges and Valve failures as well as theft and vandals as well as environmental factors etc. These hazards have so far become very common with major activities such as Drilling operations, Refinery and processing. Other critical operations include general maintenance services on Oil and gas storage and transportation facilities, e. g Confine Space and Gas line welding task, oil and gas transportation vessels, containers, tanks, underground storage facilities where a safe means of gas detection is necessary for the prevention of work personnel and assets as well as the environment from fire disasters.

Typical use of Gas Detector:

This includes:

Testing of atmosphere for combustible gas, Oxygen deficiency/enrichment and Hydrogen gas and

Responding to emergencies where combustible gas might be present.

Fundamentally to detect presence of gas leak and enrichment in gas storage areas and confined space activities.

DEFINITIONS WHEN DEALING WITH GAS DETECTOR INSTRUMENT IN CONFINE SPACE ACTIVITIES AND HAZARDOUS ENVIRONMENTS:

Flammable Atmosphere: Is one which contains adequate amount of combustible gas or vapor in the air to be flammable e. g gasoline vapor.

L E L:- Lower Explosive Limit: This is the minimum concentration of combustible gas/vapor below which ignition would not occur on contact with source of ignition. Ignition can only occur at 100% L E L [and above], any indication above this, may not cause burning since the air/fuel ratio may be too lean.

U E L: Upper Explosive Limit: Is the one which contains the maximum concentration of combustible gas/vapor in air to be flammable when exposed to ignition source. Any above **U E L** will not burn, as the maximum will be too rich. It should be noted that our gas detector could only be use to measure percentage of L E L. The indicator may not give any reading if the vapor is highly concentrated.

Flammable Range: This is the difference between the lower and Upper explosive limits; this is expressed in terms of percentage of combustible gas in air by volume.

CARE AND CAUTIONS TO BE TAKEN WHEN USING GAS DETECTOR INSTRUMENT:

This instrument is used generally for detection of low and over rich gases it is often used in hazardous atmosphere, confined space activities and gas storage facilities to detect oxygen deficiency and presence of dangerous gases that may result into fire and explosions.

DO'S AND DON'TS:

Before leaving the office, turn on instrument for a minute, then set L E L to zero position. Check battery [if poor pick up another]. Confirm that the instrument has been calibrated within the last 7 days.

- Do not use detector with poor battery level
- Do not adjust instrument while in operation
- Do not use instrument in inert atmosphere, [or else faulty reading would be obtained
- Do not dump instrument on anything while testing for gas, else faulty would be obtained
- Do not place sampling hose in liquid, steam or water. These will put the indicator out of service
- Do not use indicator to test vapor from heated combustible liquid
- Once you encounter problem with instrument in the field, discontinue activities immediately until the problem is solved.
- Consult manufacturer's manual.

CHAPTER SIX

The following questions should be asked during confined space and welding .

IN OTHER TO PREVENT FIRE:

Ensure there is permit for the task
> Possibility of hydrocarbons
> Is simultaneous operations required
> Restricted area?
> Distance from restricted area
> Is hot work likely within 35 ft/meter
> Fire extinguisher on site
> Fire watch required
> Combustible gas monitoring required?
> 02 monitoring required for Confined Space Hot Work
> Welding equipments should be properly maintained.
> Warning Signs should be posted clearly to warn against flammables etc.

RESTRICTED AREA IN HOT WORK task:

Restricted area can be described as any offshore structure that produce, process, or store hydrocarbons e. g all offshore location except living quarter structures] or any on shore location within 100 feet of hydrocarbon handling or storage facilities. Hot work shall be eliminated or minimized in restricted areas.

A SAFE DESIGNATED WELDING AREA:

A land location that does not contain or is not within 100 feet of hydrocarbon handling or storing facilities, or designated area on an offshore platform approved by the area operation superintendent and supervisor. Designated safe welding area shall become restricted area and the Hot Work Permit will be revoked if hydrocarbons are introduced into the area.

Confined space

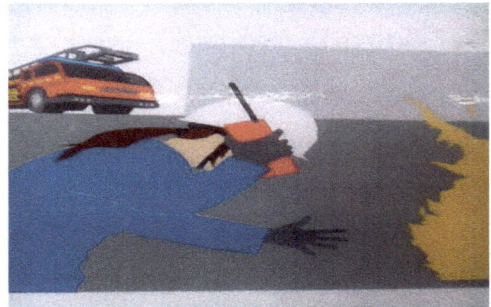

IN CASE OF FIRE CALL

SAFETY WATCH IN PIPE LINE AND CONFINED SPACE WELDING ACTIVITIES:

WHAT IS A CONFINED SPACE?

First of all let us look in to what confined space is all about based on the emphasis focused on confined space and pipe line hazards which has been a major hazardous working atmosphere when especially hot works e. g welding and grinding activities is carried out.

84

Definition:

A confined space is a space that is large enough and so configured that an employee or a worker can boldly enter and perform assigned task or work, it has limited or restricted means of entry or exit, and is not designed for continual occupancy. Additionally, a confined space may have the potential to become hazardous. Entry in to confined space introduces hazards such as physical entrapment and exposure to flammable/explosive and toxic gases and vapor which could easily result in oxygen deficiency or asphyxiation when these poisonous gases are eventually inhaled.

Example of confined space:
These include;
- Tankers
- Containers
- locked vacuum spaces
- Pipe lines
- Vessels
- Oil and Gas storage facilities
- Enclosed room
- Vault
- Tunnel

There is great need to be conversant with the hazard that confined space may contain. This include mainly of
- Oxygen Deficiency: Air contains about 21% oxygen but a minimum of 19.5% is required for entry into any confined space without breathing apparatus.

OSHA states that at about:

>14% oxygen – there is difficulty in breathing

>12% - oxygen – there is difficulty in thinking

>10% - oxygen – there is unconsciousness

>8% - oxygen – possible death

An excess or too reach of oxygen can also result in fire hazard.

Flammable or Explosive Gases: Some of these flammables / explosive gases are lighter than air and such will collect at the top of the space, while others may be heavier than air and as such will settle at the bottom of the space. Common flammable and explosive gases include carbon monoxide, natural gas, hydrogen sulfide, and methane. Methane is odorless and colorless.

Toxic Gases and Vapor: These gases and vapor can cause injury or death in low concentration and can be categorized into irritants and asphyxiates.

>Irritants are gases that in low concentration may only cause irritation to the respiratory system such as sore throat or coughing. High concentration can however cause serious concern or even death. Asphyxiates results in death.

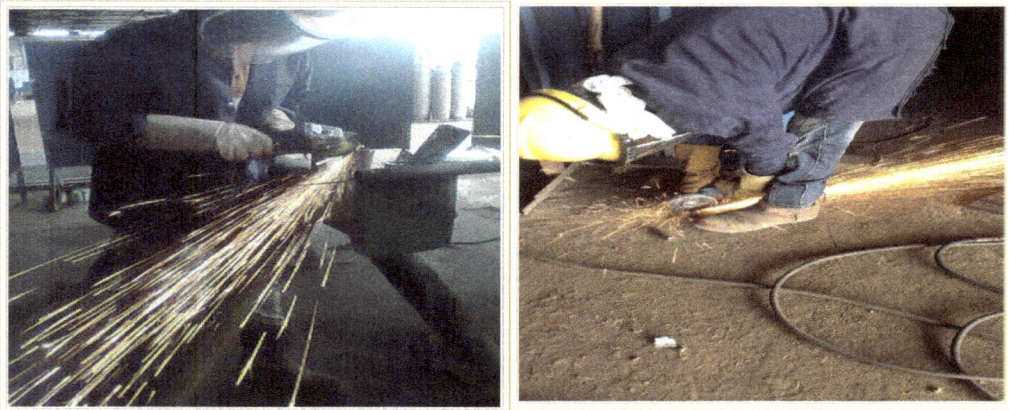

Red hot particles or sparks generated from rotating disc in grinding operations.

CHAPTER SEVEN

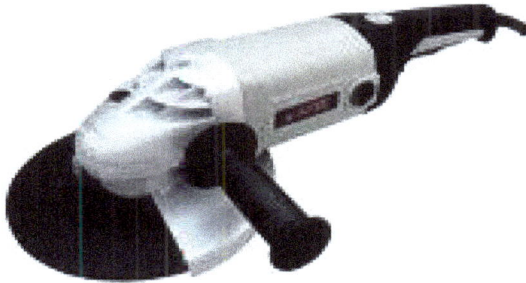

GRINDING AND FILING FIRES:

Grinding is one most critical and dangerous task that is often carried out in the industries, welding workshops and construction sites by specially trained personnel called Grinders. Grinding is the process of using grinding machine to provide a smooth and finish surface on welded metal joints as well as used to achieve other desired shapes on metal piece with the use of grinding machine as shown above. These machines are electrically operated machines with rotating fiber disc usually attached to a spindle which drives the disc to achieve smooth metal surfaces. Common hazards associated with grinding operations include rotating disc of the machine, which can as well accidentally fly out of the spindle during use if the bolt and nut holding it are not properly tightened or when they are loosed tightened and unknown to the user as well as poor quality disc type. The use of the disc in course of executing grinding task is usually

87

accompanied with the production of hot flying metallic particles that is usually reddish in color and capable of causing burns and fire when exposed to flammable substances or combustible materials within a work area. Other effects of the hot particles include burns when task is carried out without protective clothing. The hot metallic particles usually travels in distance according the power rating of the machine and increase in speed by the user which is usually determined by current supply in driven the grinder, the speed range of the machine is usually obtained at different angles due to rotational difference of the grinding disc which imposes a high possibility of imbalance which can result to cut and fire through the disc when placed under extreme pressure on metallic work piece in close proximity with combustible material.

Therefore, serious care is required when using grinding machine because of the potential hazards involved in the task.

Cutting with Rotating Disc & Oxygen-Acetylene Gas System:

FIRES FROM OXY-ACETYLENE CUTTING:

Cutting is the process of using oxy-acetylene combustion process in cutting of iron metal piece to obtain different shapes and sizes. This activity is most often carried out in the industries, welding and fabrication workshops as well as construction sites. There are lots of inherent dangers associated with the use of the process in cutting task. Some of these hazards include; Direct cut injury from the direct heat produced from the burning nozzle of the Oxy-Acetylene cylinders, entangling with power cables especially when cutting machine is in motion and when the cutting machine is not properly handled upon which sparks or hot slag can as well cause severe burns. The sparks or slag produced by this machine is usually a red hot glowing particles and often in a

89

random motions during cutting operations. Fire can occur when these particles becomes exposed to flammable substances as well as can cause burns where protective PPE and protective barriers are not provided to prevent spread and travels. Other hazards include possible rapture of the hose insulations by heat which harbors the gas line when they are not properly handled. Several cases of industrial fire outbreaks have resulted through inadequate maintenance of these devices such as explosion from defective valve and flash back arrestors which controls the regulation of gas movement and pressure from the hose line to the cylinder as well as when inadequate attention is not given especially during grinding and when the task is carried out in close proximity with combustible substance and materials at work place.

Defective electrical appliance & maintenance fires:

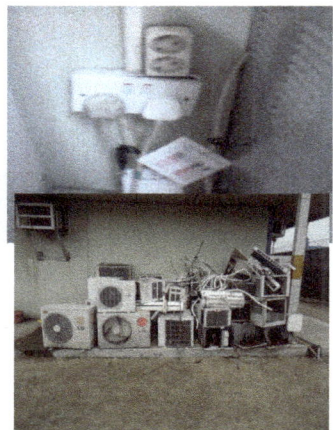

Today the uses of electricity have became very imperative to human due to its numerous advantages over other sources of energy in driving domestic and industrials equipment and appliances. Industries, construction sites, workshops, homes, salons, offices and all business operated outfits are all in use of electricity because of the importance of electric power. Electricity is used for generation of light, powering of industrial equipments, such as showing machines, electronic and electrical appliances, electric motors, power hand tools, welding works, production as well as office use such as Air Conditioners, water dispensers, lamination etc. Today the use of electricity in the operation of the above industrial equipments, machines, home use and office equipments have so much helped in the enhancement of business and production activities both in the private and the public sectors than when activities were manually carried out without the support of electricity in the old. However despite the intervention of electricity and the benefits to society yet, there are still lots of potential hazards associated with the use of electricity both at homes, industry, construction sites and offices.

What are the dangers of Electricity?
These hazards include;
>Electrocution/shock
>Fire
>Explosion
>Radiation effects
>Induce thunder fall

What are the Common Causes of Electrical Problem?
All the above mentioned hazards can easily come to occur especially when electrical installations and connections are been wrongly made, including wrong application of electrical

appliances and equipments by people. Others include the use of substandard electrical materials in electrification and installations works.   as well as the use of incompetent personnel in electrical installation works, and applications of defective electrical appliances and fittings e. g Air conditioners, Freezers, Irons, use of low quality standard electrical materials in installation and repair works, wrong electrical connections, unearthed electrical connections and wiring, using electrical appliance under rain or wet surfaces, over loading of electrical appliance and equipments, short circuit, taking short, by passing, not using the recommended capacity of cable and wires for the right load or equipment, improper insulation of electrical connections, sparks from poorly tightened cable connections, placement of low voltage electrical appliance/equipments on high voltage output, improper maintenance of electrical power generating sets and placement of combustible materials too close to ignition sources e. g petrol and naked flames or sparks, defective connections and poor contacts from power transformer cables, over head electrical line in contact with metallic materials, improper tensioning of power lines resulting in contacts with heavy duty trucks and equipment during transit along major roads, rough and unprotected connections within houses, over loading of cables due to house to house connections, over loading of public transformers etc. All the above violations can actually result to either overheating as well as sparks which can eventually result to major fire outbreak.

Precautions and Control measures against electrical fires:

- Before commencement of any electrical works obtain authorized permit. The permit will ensure that the work is carried out in a safe manner.
- All electrical installation and repair works must be done by competent personnel.
- Ensure that all electrical works are carried out in line with regulatory requirement and standard.
- Ensure adequate assessment of work area before electrification task is carried out.
- Ensure the use of right tools while executing electrification works
- Ensure the installation of thunder protection devices at work place, homes and offices to protect against thunder strike and fires.
- Ensure the use of recommended electrical materials in all electrical installation works.

MORE CONTROLS INCLUDE THE FOLLOWINGS:

Use of poor electrical appliance and equipments at work place as well as non compliance to safe work procedures should be avoided. However, there are basic recommended measures on how best electrical works can be done safely both in the industry and construction site and homes and is usually provided and detailed to be followed accordingly especially during installations, connections and use of industrial electrical appliance and equipments such as power hand tools such as Grinding Machine, Drilling Machine, Saw, cutters, electric iron benders, electrical filing machine, air conditioner, refrigerator, heaters and cookers etc. Some of the control measures include the use of Check list on all electrical appliances and installations as well as inspections procedures which is usually conducted by competent personnel to meet requirement and to identify possible defective electrical situations to provide corrective measures for safety. Others include use of warning signs such as Warning tags on defective power tool, labeling of switches and circuit breakers, placement of maintenance log book, list of electrical equipments, log book for power tools etc.

OIL AND GAS FACILITY MAINTENANCE FIRE:

Oil facility is a location where assemblies of mechanical, electrical, electronic devices and component units with associated civil, piping and instrumentation installations are integrated in a field with the capability of these facilities to control, meter and monitor all oil flow pressure, quantity

94

measurement and transmission performance of oil and gas storage with control systems to enable operators have effective control over petroleum exploration production and utility operations in the oil and gas sector. These facilities may however have the storage of the following associated gases which include the storage of non refined gases, natural gas, crude oil, liquefied gas etc in a crude or refined form as the case may be with a highly classified potential risk of fire and possible explosion during operations due to the nature of sensitivity of this facility.

Some of the associated fire risk and hazards in the oil and gas storage facility operations include the followings:-

Fire during maintenance works e. g .spark/ignition through friction and mechanical tools striking against one another
Explosion
Oil spill
Oil theft
Illegal Bunkering/vandals
Use of explosive devices
lectrical
Gas leak
Smoking
Hot works e. g welding in confine space and gas storage area without proper isolation and control
Extreme surface heat [heat radiated from sun light]
Fire during maintenance works [mechanical or electrical sources]
Arson/enemies
Direct flame e. g smoking

All the above activities possessed their individual potential fire risk and other forms of actions that can result to spill, leak, spark and fire when they are inadequately carried out within an oil facility /locations or when proper prevention, protection, controls, isolation, elimination, use of recommended tools, safe work procedures and use of competent personnel are not been provided to ensure guarantee of safety.

OIL SPILLS WILD FIRES:
Introduction:

Oil spill is an accidental discharge of oil from pipeline or facilities. It can pollute land and water as well as cause great fires depending on the nature and how as well as where it occurs. Oil spillage has caused several environmental degradation and major damages several oil and gas facilities, such as installations, pipelines, plat forms and manifolds, storage tanks and well as bush and farm lands, creeks, rivers, camps and sometimes homes with great lost to lives as well as disabilities and displacement of people and families, communities, towns and governments with most times great impacts on the natural environment as the case may be.

96

Several oil spills fires have resulted in massive wild fires with great destructions to lives and properties and major cities affected from its disasters depending on the environment that it occurs.

Causes of oil spill fires are many but will include the following:
Sources of Oil Spill Incidents:

These include:
Equipment Failures e. g valve, pump, control unit.
Human error
Corrosion
Sabotage/theft
Drilling operation
Emergency operation
Others e. g. construction, natural effect e. g. earthquake etc.

Main causes:
SPILL CAUSATION, ACTIVITIES/FACILITIES
Equipment Failure- e. g. seal failure etc.

Human error- e. g. Operation and maintenance technicians

Corrosion- e. g. Flow lines, delivery lines, fittings etc.

Drilling & Engineering- e. g. dredging, pipeline replacement

Sabotage/Theft- e.g. Willful damage of oil production facilities e. g. flow lines, manifolds, tampering with gas lines.

Natural Causes- e. g. flooding, heavy rainfall, falling trees, lightening, etc.

OTHER EFFECTS OF OIL SPILL FIRE INCLUDE:
Community unrest from the outbreak/ fear of fire and disruptions
Death of aquatic life/ causes of hunger/ economic lost
Stunted growth in vegetation/ causes of hunger/ economic lost
Disrupt recreational activities/ economic lost.

All the above are other effect of oil spillage but fire outbreak is the main subject matter in this series because of the immediate impact that it can cause to man and the environment.

Strategies and controls for Oil Spill Prevention:
- **Use trained personnel**
- **Use of equipment designed to specification**
- **Regular inspection/ maintenance of equipment/ materials**
- **Surveillance of facilities**
- **Facilities upgrading**
- **Adoption of corrosion control**
- **Improved community relations [C L O]**

Conventional Oil Spill Control Methods:
Isolate/stop source
Containment e. g. Boom, Absorbent, Skimmer
Recovery e.g. Barges, Dispersants
Storage/transport
Clean up [mechanical]
Mop up [physical]
Disposal
Rehabilitation
Certification [appropriate department, DPR, Contractor, Community.

Social Imbalance as some major causes of oil spills:

Despite the above existing conventional measures been adopted in the prevention of oil and gas spills in the oil and gas industry apart from those identified challenges of theft and vandals, yet there are still lots of more hidden contributing factors to the real causes of consistent oil and gas spills disasters in oil and gas facilities. These problems should however, be viewed as managerial oversight and or a deliberate display of inconsistencies in this sector. One major issue is the problem of inadequate remuneration for oil and gas workers compared to how these categories of work force where paid back in the days. This is in consideration to especially Welders, Grinders, Non Destructive Test [NDT] Specialist, Supervisors, HSE personnel and other key workers who play active roles with sole responsibilities of caring out critical task executions as well as others who enforce monitoring and effective supervision of critical task execution to ensure flawless task delivery at sites and during facility installations and maintenance works. The major inducing causes of this problem is the fact that most of this categories of workers in this sector are still not given adequate care with very poor welfare as well as very meager salaries compared to the nature of jobs that they offer to this sector. Why this has actually become a problem can be viewed with respect to the present day situations of demands and high cost of living due to population increase where majority of the workers in this sector that could no longer coupe with this meager salaries and yet still without improvement or support. The adage which says, Hungry man is an angry man, this should also imply to how safe work can be delivered or not when workers feels unhappy and angry at work. Let us also look in to what could possibly happens when a welder may feels unhappy at work, first, they become exposed to all sorts of industrial hazards because of being unhappy while at work thus they see every aspects of their work frustrating and thus will impose them high level of potential hazards. Secondly, these conditions might easily induce them in to different kinds of thoughts, conceptions and actions that can sometimes be displayed on the job they do. But in all, no Welder or Grinder can be able to deliver a flawless task without creating an atmosphere for mistakes and errors to occur when they are not happy. Others include deliberate actions that these set of people can actually take to cause serious problem at work when they are not happy. The above analysis should be able to explain the risk and implications involved on these over sighting or inadequate remuneration for workers as the case may be. However, let us now consider the above scenarios as well to see how it goes with the oil and gas pipe welding operations. Majority of the welders today on the excuse of not been well paid have rather taken different dimensions by seen how best their lost efforts in task executions with these sectors can be technically replaced with how many welded joint they can produce in a day, no matter what it would cost the owners and operators of these facilities at different instance. Good numbers of professional welders today have inculcated the habit of caring out improper task executions on given jobs by engaging in reductions of structural integrities required of oil and gas pipe lines constructions. E.g [Issues of defective welded joints]. When these welded joints between various sizes and capacities of pipes used in the transportation of crude and gases along pipe lines, installations and or manifolds are not provided with required integrity to withstand certain environmental and atmospheric temperatures, especially when exposed to extreme hot climate definitely these pipes with their poorly welded joint would result in leakage where spills can comes to play as soon as these materials becomes subjected to various environmental conditions at different points or locations hence need for adequate remuneration for oil and gas workers should be given a fare priority to ensure adequate and safe task delivery to avoid unnecessary oil and gas spills disasters in our environment.

CHAPTER EIGHT

MARINE TRANSPORTATION & HOUSE BOAT FIRES:

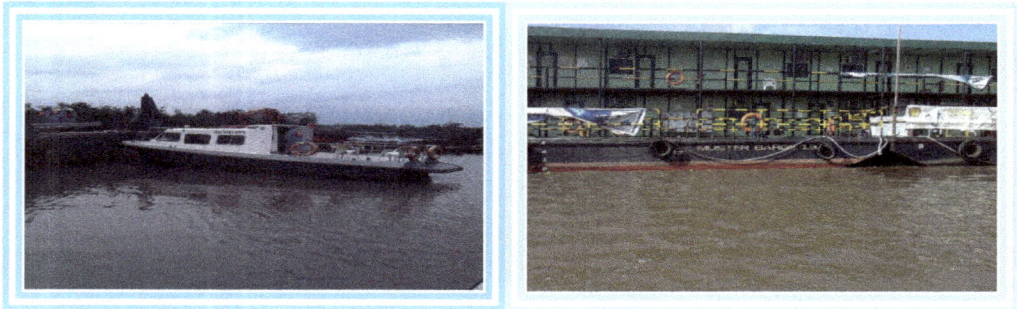

Marine transportation has remained a fundamental source of transportation especially to coastal dwellers and transportation of heavy merchandise. This route has been of major importance in the area of conveyance of goods, people and services especially in the marine trades in coastal regions like Niger Delta region of Nigeria and other parts of the world where majority of the settlement is surrounded with water. These routes have made trades very easy and have equally promoted economies of many nations especially in the area of import and exports of heavy merchandise. Different sea routes and rivers have been used in navigation and transportation of people, goods and services all over the world by people especially in the transportation of oil products such as crude and gas with great risk of fires. This means of transportation have been of great help to people in the trade and transport sectors than any other means of transportation because of the fact that several capacities of cargos and loads carried by water cannot be easily carried on land and air by vehicles and aircrafts without creating obstructions and inconvenience against traffic due to extreme weights and sizes of certain cargos. Despites the convenient and benefits of sea and rivers used as a safer route in transportation yet there are still different types of associated risk and dangers with marine transportation. These hazards include, Boat mishaps, Man-over boards including marine and vessel fire outbreaks.

COMMON CAUSES:

Common cause of these problems are mostly due to human faults, mechanical and electrical systems failures e. g. heat generated through friction from non lubricated moving parts of vessel engine, sparks from live cables contacts and battery terminals, dropped metallic objects on battery heads, defective electrical connections, live cables in contact with earth body of engine, overheating of engine, ironing of cloths and use electric iron, heaters inside house boats without adequate care and fires from combustible substances in contact with ignition sources. This include, when people fail to following or comply with marine transportation safety rules and regulations, use of incompetent sailors, drivers, beach masters etc when embarking on marine journeys. Other factors responsible in the cause of marine fires include fire from vessel accidental crash e .g. head on collision of two vessels, boats etc, arson by enemies, cooking with adulterated kerosene which has been a major cause of several marine fires including improper use of combustible materials and ignition sources e.g. during refueling of engine, transfer of fuel, fuel leakage, fuel spills, smoking, lighting of matches, lighters, use of candles and mosquito coils as well as wrong connections of engine battery terminals on marine boats, house boats, vessels, pusher trucks and other types of marine crafts. These problems have caused many marine fires and resulted in several property lost and death especially in the coastal areas where marine transportation is in common operations.

OIL BUNKERING AND VANDAL FIRES:

Oil theft such as bunkering and vandals of oil pipe lines have remained major causes of environmental pollutions and real causes of several industrial fires. These fires occurs when people illegally make deliberate attempt by tampering with crude oil pipe lines and installations for sole purpose of tapping crude oil and allied products for sales. The conducts and actions by people in this direction have seriously caused several havoc and resulted in several massive pipe lines and installation fires with great lost from repair and maintenance cost on individuals and government as well as controllers/operators of these facilities with extreme damages and effects from these acts, and other existing hazards caused along these crude lines. Crude line fire occurs when vandals and theft personnel during process of cutting of these pipe lines imposes high possibility of fire to occur because of the presence of crude pressure flowing along these lines, the processes of introducing extreme heat to the line by cutting with oxygen/acetylene gas energy in a way of gaining access to the crude reservoirs, the temperature of the pressurized crude gas is raised to their flash point where spills and fire is eventually generated. This situation is usually worse when oil lines are tapped with the use of cutting process either with use of oxy-acetylene gas cutting or other dangerous process such as direct explosions. These problems can occur anywhere around the globe depending on where bunkers and vandal may target as their agenda. This is one situation that have also induced several marine fires from direct oil spills pressures in contacts with naked and unprotected engines of marine boats and vessels as the case may be, other common causes include when loaded crude vessels are sometimes ignited due to poor coordination of flammables and ignition sources as well as fires by arson and pipe line leakage and equipment failures in presence of thunder lighting or other direct source of light with sometimes open gun fire attacks between government security operatives/surveillance team and vandals in open battle against vandals and and crude oil theft in the creeks and other routes where oil pipelines and facilities are crossed or installed especially in the oil and gas producing regions.

PREVENTIVE MEASURES:

- Adequate preventive measures should be enforced by government by incorporating community vigilante in monitoring of oil pipe lines to avoid vandal and thefts.
- Before embarking on any marine journeys, journey managers and beach masters should ensure that proper checks is be carried out concerning how flammable materials and hazardous substances are loaded and carried in their boats and vessels.
- Boats and vessels should be accompanied with firefighting equipments such as fire extinguishers etc, including qualified firemen to be on board.
- Boat drivers and deckhands should be well trained and competent in marine transportation operations.
- Smoking should not be allowed specially closed to engine compartment or flammable substances
- Always ensure that good housekeeping is maintained on board at all times
- Drivers and passengers should always be aware of boat mishaps and accidental spills of petroleum products.
- Caution and warning signs of fire hazards should be visually displayed on board during any marine journey. e. g **NO SMOKING, NO NAKED FLAME, etc.**

Always ensure that good electrical connections are maintained by competent personnel on board and ensure that engine is protected.

Ensure maintenance works are carried out by only competent and authorized personnel at all times.

Transporters and passengers should be aware of accidental fire outbreak at any point in time as far as marine journey is concern this can occur from exploded oil vessels by government security personnel in combat against oil theft at sea etc.

Finally safe marine journey management rules and regulations should be enforced and complied by all at all times.

Caution should be maintained at all times by marine operators on possible oil spills incident and there hazards when embarking on any coastal routes journeys.

FIRES FROM POOR WASTE MANAGEMENT:

Lack of waste management today have become a source of many environmental challenges, this include land and water pollution and degradation of the natural environment. Causes of these problems are results from lack of adequate waste disposal and management resulting from inadequate enforcement of legislations and actions against the ill practices. Poor waste disposal has been a common practice especially in many parts of the world from time until now that the impact is felt. This is widely practiced and very common in the rural and even common in some urban areas. Several impacts from poor waste management have been experienced today by people this includes air, land and water pollutions as well as domestic and spontaneous fire outbreaks as s environment ome of them. Hazards of poor waste management can as well cause fire out break both at homes and work place especially in an industrially operated and construction sites or where effective waste management is not practiced. This challenge is due to how different types of hazardous materials and chemical substance are used in manufacture, manipulation of products as well as execution of work processes in certain industrial activities as well as how they are been disposed. The uses of these chemical and hazardous substances in the manufacture and construction works both in the industry and construction sites without effective segregation in to their waste streams and adequate disposal their adverse reactions with other environmental elements are causing several havocs and resulting in several environmental damages including spontaneous fires at especially in hot climatic regions. Most causes of these fires are due to how chemical substance waste and others are been disposed off without adequate consideration to their chemical natures, harmful effects and reactions when they are poorly disposed off especially when disposed within certain environment. Some of these hazardous or toxic wastes include hospital and medical waste, radioactive material waste, electrical material waste, gaseous waste, which are all chemical wastes etc. When these chemical wastes are generated and disposed off at certain environment without adequate segregation of their waste streams from others it seriously becomes hazardous and forms a vacuum for various reactions to take place because of their individual chemical properties. When these wastes are dumped together with their individual chemical constituents without separation a reaction takes place and the resultant can induces spontaneous fire occurs. Spontaneous action takes place when these chemical reactions becomes exposed or subjected to certain environmental factors such as extreme heat from sunlight [heat radiation from the sun] and other direct source of heat processes. Because of the dangerous natures of these wastes some approved conventional measures for effective waste management and disposal has been developed in to use for public safety. These measures are adopted and used both at homes and especially in the industries and construction sites to minimize or reduce their impacts to man and the environment.

LANDFILL TYPES:

CHEMICAL LANDFILL:- Waste requires pre-treatment before disposal-hazardous industrial waste.

SANITARY LANDFILL:- Biodegradable materials, municipal waste/ industrial and non-hazardous waste.

INERT LANDFILL:- Non decomposable, non- water soluble wastes e. g. nuclear wastes. Buried deep in the earth to about 97, 500 m] and monitored.

OTHER HAZARDS OF POOR WASTE MANAGEMENT INCLUDE

- Unaesthetic dump site
- Foul odors - loss of community pride
- leach ate from dump sites can poison our surface and ground water.
- Stagnant pools provide breeding ground for mosquitoes, flies and other disease vector.
- Provide abundant food for rodents which transmit harmful bacteria and virus leading to epidemic – Ebola, Lassa, Plague health problems and fire outbreak.

CAUSES OF MARKET FIRES:

Market is where buying and selling of commodities in exchange of money takes place, market is usually a designated place sometimes established by Government, communities and individuals usually with buildings and structures such as stalls, shops and sometimes wooden and zinc batchers and attachments where people use as shelters to provide protection for themselves and their goods while in business transactions, also some of these structures are also for storages of different types of goods and commodities to avoid theft and damage. These categories of goods include solid, liquid and gaseous products which are all combustible in nature as well as others that are non combustible at normal temperatures. Most of these structures that are built at market places for purpose of safety, security and convenience to provide safe atmosphere for business operations with most of them built combustible materials that can easily be ignited when placed under certain atmospheric conditions. However, today cases of market fire have become very common and occurring almost day to day due to sometimes the way combustible materials are been handled in close proximity with ignition sources as well as fires deliberately caused by arson. Some common causes of market fires include, how stalls, shops as well as batchers and attachments are build and attached closely to one another in their closed proximities with most times market stalls roofed together without due consideration to possible fires. Another major issue is the high level of ignorance amongst market traders especially market women on their ignorance about common causes of fire and how fire do spread from one point to the other if eventually there is real fire outbreak. Lots of market fires have been traced to causes from poor handling of combustible materials and substances as well as electrical fires that are often results from poor electrical connections, e.g use of low rated electric cable and wires in wiring installations at stalls with many stalls most times observed haphazard and un-insulated naked connections with less regard to their implications, use of defective electrical appliance, overloading of wall sockets cables wires, due to often times with these stalls and shops connecting light from one another. Other major causes of market fire in include accidental fuel spills fire from the locally called One Man Carrier Generators during refueling; the use of this portable power generating set becomes very common at the market due to incessant power failures in different parts of the world. Common causes of its fire include accidental contacts between two naked cables resulting from poor connections, ignition sources coming in contact with flammable substances, careless smoking of cigarette, Cooking with defective cooking stoves, poor handling of cooking gas cylinder bottles, poor use of matches and candles during power outage, including use of mosquito coils at market places sometimes by children and security personnel during security operations at market places etc.

Control Measures:

Market stalls should be build and provide with enough space from each other to avoid easy spread of fires in event of fire outbreak.

All electrical installations should be carried out by competent personnel to avoid fires from wrong electrical connections.

All marketers, traders, buyers and sellers, occupants of markets stalls and shops should all be adequately enlightened and educated on common causes as well as prevention of market fires.

Traders should be well enlightened to avoid unsafe acts and conducts that can easily induce fire at market place.

There should be strict and consistent monitoring on how combustible materials and substances are stored and moved by people at market places.

Sources of naked flames should be strictly avoided closed to where combustible substances are stored at the market places.

Barricade or caution tapes and other forms of physically protective barriers should always be provided and used to caution people of presence of combustible substances against unexpected market fire.

Fire fighting equipments such as fire extinguishers of all types and fire hydrants should be made available and placed visually at strategic locations at market places where people can easily gain access whenever there is fire.

There should always be provision of adequate access roads in the construction of markets places including adequate escape routes so as to enable adequate control of people movement and evacuation as well as to pave safe way for fire fighters in the event of any market fire emergency situations. Several market fires have occurred in different parts of the world and resulted in serious destruction to lives and properties worth millions of dollars because of inadequate access roads and escape routes to provide means for people including fire fighters who may come in rescue and combat of such fire situations. The sole advice to all market traders all over the world is to see prevention as the bed rock to prevention of all classes of accident including fire which have destroyed and claimed several lives in different parts of the world. Hence for traders to always imbibe safe handling practices on all combustible materials in compliance with regard to safe handling procedures to avoid risk of great fires at market places.

TIME FACTOR AND ACCESS CONTROL IN FIRE FIGHTING:

The most important factor in fire fighting is time factor. Fire discovered on time can be quickly put off. If fire is observed on time and responded quickly fire can be promptly put off, because whenever fire is delayed it becomes a problem and create a platform for spread. Any delay in discovering or fighting fire would create avenue for fire to rapidly spread from one start point to another. This is the stage known as incipient to inferno stages in fire outbreak]. Many market fires have occurred and caused huge destruction to lives and properties due to most times delays in discovering such fires as well delays in fighting such fires as well as sometimes due to lack of fire men, non availability of fire fighting equipments including lack of good access roads and other forms of obstructions during major fire outbreak especially at market places, Gas and Oil storage facilities and other major construction sites. Fire teams or fighters have faced several great hindrances during major fire outbreaks because of inadequate access road net work. Fire fighters have always found it very difficult in many cases of fir outbreak to gain access of their ways in to most fire scenes because of inadequate access roads and obstructions usually created by people. There are several areas where houses are built in a more congested nature with obstructions where fire fighters have always found it impossible to penetrate even in major fire outbreaks. Other factors include poor communication during real fire outbreak, the way and manner at which information is passed to fire fighting agencies/firemen have also help in causing more havoc than doing good in fire emergency situations.

CONTROL MEASURES:

Communication need to be given a priority and should always be in a straight and concise manner and central when dealing with fire emergency situations. Any information wrongly passed in any emergency situation can as well frustrate and jeopardized whole efforts and cause more havoc and possible deaths. However time factor should be considered more paramount at all times, whenever fire alarm is raised or observed, priority should be given on how alarm is raised, how people should respond to emergency information and situations including evacuation as well as how firefighting equipments should be mobilized and provided access for used. Provision of good and adequate access to fire fighters during fire out breaks e. g. how fire trucks are mobilized and given access and support without hindrance or further obstructions by people at fire scene in real fire outbreak plays a major role in the success of every fire fighting.

MAJOR CAUSES OF FIRE SPREAD:

Scenario of fire out break with how fire can spread from one fire point to the other just as shown with the above filling station fire scenario.

Causes of several filling station fires are results due to closed proximities between petrol stations with respect to how residential buildings are also been erected at different locations, as well as how markets are also built in different parts of our societies and more especially at densely populated area with congested residential buildings without due consideration given to those associated potential fire hazards with these areas. Today it has been observed and a common practice where you see various categories of houses build very close to filling stations in different parts of our societies on the fact that most places where these houses are build are just on premise of economic and hardship reasons. Most of the houses been erected close to filling stations are structures built because of the majority of these owners left without options, and some with deliberate tendencies to make money without any course of verifying possible hazards and unforeseen calamities that may arise as the case may be. It has also been found at several places where

117

markets and filling stations as well as residential buildings have already been erected in close proximity to one another yet, the question have remains on how best this situations should be addressed as in regard to associated hazards with petroleum business and residents. In many occasions the question had remained who could have first settled on such land. Critical situations of these natures had also been dragged to court by some residents in view of challenging these contentions. So far in many of such occasions courts in their fundamentals have been able to adjudge many litigations in favors with regards to first settlers on such lands. The question now is should government or regulations allows individuals to sell their individual lands to filling station owners upon violation in the part of the petrol dealers on individual accounts, and or should government still maintain existing restrictions against filling station owners not to built filling stations closed to residential areas which have been a mandatory regulation by Department of Petroleum Resources in regulation of petroleum activities?. However, there are still many other factors that are still required to be considered in respect to why lots of buildings are still been erected closed to filling stations or the reverse as the case may be. The issues of economic factors is another critical situation which requires serious correction and amendment at all levels this is because of the fact that majority of the land owners along major high ways and towns where filling stations are built in close vicinities with residential areas are categories of persons that do not actually have the intensions of allowing filling stations to be erected in close proximity to their houses considering associated hazards but it is an issue arising from several economic situations that have made some of these land owners through influenced of money been lured in to accepting offers from petrol dealers by allowing petrol stations built within their territories despites associated risk involved in petroleum business. Today incidents of filling station fires have become very common in different parts of the globe because of the associated hazards in close vicinity with residential areas and the business even with majority of the land owners who have ignorantly allowed filling station owners to built close to their houses in serious regretting modes especially whenever incidents of filling station fires is heard. Filling station fires have actually occurred in different places and resulted in several property destruction and deaths with adverse impacts of disability and displacement of families yet with no definite solutions to these challenges. Some of the challenges associated with filling stations include fear of immediate fire spreads which is often induced through radiation from one fire point to the other. Fire spreads are usually induced by the nature of fire and the type of material substance that constitute actual burning of the fire and according to the quantity of heat generated as well as atmospheric conditions which is temperature at the time of fire and other mediums which can easily help to facilitate transfer of heat from one point to the other such as wind and time factors. Heat is most often transferred from one body to the other through radiation, convection and conductional means. However, let us look at few definitions and meaning of the above processes when dealing with fire as the case may be.

CONDUCTION:

Heat transfer from the area of high potential to a point of low potential.

CONVENTION:

Heat transfers through heated gases and water. [Bush burning]

RADIATION:

Fire can be caused by radiation when the heat reaches ignition temperature at the other end.

DIRECT CONTACT:

Direct contact is when the burning object or the heat source is in direct or close contact with the surrounding materials. When temperature of the close or surrounding material becomes raised fire will occur, e. g. lighting candle with matches.

Wind as a medium of heat transfer has been responsible for several untraceable fires in many parts of our societies. This happens when already hot or heated particles [ashes] in their glowing conditions are been moved in to the air through the help of wind and transferred from one burning point to another. This principle implies to how birds ate seeds and take it to other places. These hot glowing particle ashes when dropped on any dry grass land area will automatically generate fire and eventually give rise to the spread of such fire from one part of the grass land to another. The above analysis provides similar actions to how fire is been spread even at market place through radiation and conduction.   Most markets are been provided and installed with electricity to provide enabling environment for traders to sell electronics and other electrical appliances, also to enable them carryout test running of purchased items by customers for items status confirmation at the markets. However the introduction of electricity in to the market place for purpose of security and convenient to an extent have placed many markets and shop owners under serous hazards. Many incidents of fires have been recorded with most times caused by people due to ignorance and carelessness, and also how combustible materials and ignition sources are been stored and used at the market place. Numerous cases of great fires and property damage have been recorded from market fires upon which huge amount of money and lives have been lost. Most fires commonly encountered at market place include solid, liquid, gaseous and electrical fires:

Some of the causes include;

Smoking in presence of combustible materials

Poor handling of combustible materials in the presence of naked light or ignition sources at the market place

Poor electrical connections

Use of defective electrical appliances

use of low voltage electrical cables in place of recommended types

Use of incompetent electricians for installations work

Over loading of electrical cable with too many appliance at a time

Use of electrical heaters, cookers, iron, and other accessories without care.

During transfer of petroleum products from one can to another

Use of portable power generators indoors

Use of candle and mosquito coils at night.

Use of one man carrier generators indoor

Arson etc.

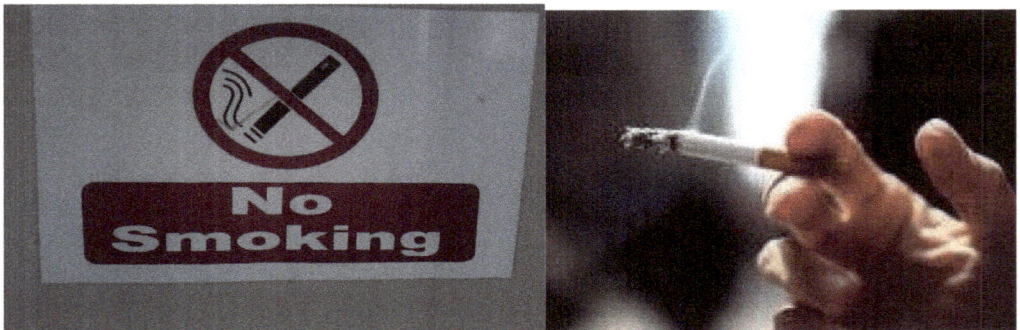

CAUSES OF FIRE FROM SMOKING

Smoking has been a major cause of most domestic and industrial fires. Smoking generally at homes, public places, fuel and gas storage areas as well as workshops have become major causes of public, domestic and residential areas as well as vehicle and industrial fires. Fires that are commonly caused at these places are usually caused due to people's carelessness. This happens when people sometimes undermined the risk of fire and smoke within the storage of flammable materials or substances, e. g. Within petrol and gas storage areas, Poor handling of combustibles materials around smoking areas and smoking during maintenance and construction works by repair technicians at work shops and construction sites as well as smoking while caring out hot work activities e. g during welding operations such as welding in confine spaces, repair works at filling stations, oil facilities, automobile workshops, chemical laboratories, plants and fuel depots etc.

SMOKING A MAJOR CAUSE OF FIRE:
The 12 Life Saving Rule No. 8

Globally, causes of fires had been attributed to several conducts and action by people with less percentage on natural occurrence. However, one commonest factor which has been identified as a major cause of fire is the conventionally acceptability on smoking as well as how petroleum products are been used, handled and stored especially during refueling of vehicle and maintenance at workshops and garages including filling station activities. Smoking in public places, homes and during vehicle maintenance at garages and workshops have actually resulted in many fire outbreaks with property damage, vehicle burnt, injury and loss of life as the case may be. These situations have given rise to the use of fire fighting equipments such as fire extinguishers and other systems including the introduction of the current life saving rules in enforcement at work place in other to provide adequate controls and prevention of fire at homes, offices, and work places. Some times for purpose of prevention, the following signs are always displayed both at maintenance and construction areas including filling stations to warn the public against fire hazards.

These include,

>No smoking caution signs.

>Do not smoke outside designated area etc.

>Do not use mobile phone

>No naked flames

>Switch off vehicle engine

>No direct flames etc.

All the above situations can occur and give rise to petroleum fires which is one important need for people to be fully conscious and aware of the associated hazards during purchase, use, storage and maintenance activities. These causes have given risen to many fire situations mainly due to human factors which include attitude, ignorance, common mistakes, carelessness, and negligence to safe safety precautionary measures provided for the safe handling of all combustible materials as recommended by laws.

SOME CONTROL MEASURES:

>Vehicles to be repaired or maintained should only be handled by trained and competent hands and to be

maintained in a safe and conducive work environment spacious enough to accommodate all repairs activities free from heat and combustible sources.

>Smoking should be avoided during repair maintenance works.

>If for any reason a repair personnel should smoke such workshop should be provided with designated smoking area outside the repair maintenance zone as well as accompanied with fire extinguishers and fire blanket.

>Naked light and combustible substance should be kept and stored very far away from repair areas.

>Workshops should be provided and installed with fire fighting equipments such as smoke detectors, fire hydrants including provision of adequate waste storage and disposal facilities.

>Waste generated at workshops should be separated according to their waste categories and timely evacuated to avoid spreads of infection and spontaneous fires from chemical reactions.

CHAPTER NINE

CAUSES OF AUTOMOBILE MAINTENANCE FIRES:

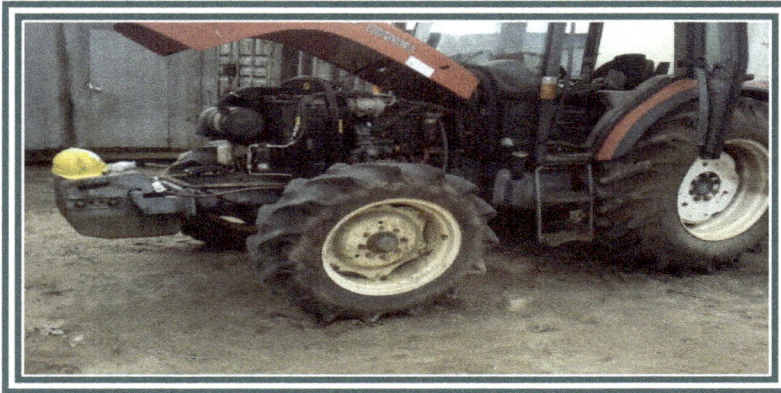

Vehicles are used for conveyance of people, goods and services. The use of motor vehicle in the facilitation of business and services today has played very important roles especially in the area of mobility and business transactions. But today the story has become different with the fact that almost every day we hear cases of induced automobile accidents with many resulted from fire outbreak and explosions. Occurrence of vehicle accidents especially fire cases has become something else due to the nature of associated risk and dangers involved in the operations. These problems have in frequent occurrence because of how cars and heavy duty trucks are been used on the roads especially **fuel tankers,** including how **maintenance work**s is been carried out by users and technicians. These challenges has often occurred due to sometimes inadequate maintenance attention given to vehicles by users and owners for not been able to provide required treatment, and items for cars and vehicle as at when due, use of second hand spare parts, defective spares, importantly the use of wrong hands [incompetent personnel] in maintenance as well as use of untrained and incompetent drivers in transportation operations as the case may be. So many other factors are responsible as causes of vehicle fires upon which many crashes have been encountered by drivers such as direct vehicle collision **from poor driving skills** and total vehicle burnt down of vehicle either through contacts of fuel spills and ignition sources. Other causes include road racing by co- drivers especially fuel tanker drivers. These acts have induced and caused several automobile fires on several high ways. Causes of these include fuel spills from fuel tankers in contact with direct flame or ignition sources, poor electrical wiring resulting to short circuits and sparks, crashes due to most times bad roads causing falls of fuel trucks etc, poor lighting, defective visions or visual challenges of some drivers, conveyance of petroleum products at night, during refueling as well as how combustible materials are used near ignition sources, e. g fuel leak in presence of hot exhaust, electrical sparks from vehicle cables in the presence of fuel leakage, sparks from loosed battery terminals and cables, wrong wiring or connection by incompetent repair personnel including smoking by maintenance personnel and drivers during vehicle repairs and accidental discharge of petrol [spill] from pump nozzles through distractions at filling stations. However where this problem occurs mostly is during mechanical and electrical maintenance services at workshops and garages and exposure of petrol tanks to extreme heat or hot surfaces e. g. during welding as the case may be.

The issue of electrical challenges occurs when electrical connections of cable and wires interface with one another in the presence of combustible materials. Others include arson by enemies and fires from static energy through friction by human body and car body in contact with fuel vapor and electromagnetic radiations in the presence of fuel spills and vapor, leaks from fuel Cans carried inside vehicle boots, during transportation of petrol products from depots and filling stations. Others include distraction which is usually caused by mobile phone when they are used during sales of petrol to motorist at petrol stations. Several auto mobile fires have happened in many parts of the world without actual trace to their real causes therefore the need for car owners and users as well as car maintenance technicians including filing station attendants to be fully aware of these hazards. The above information should be able to demonstrate and analyzed some of the common causes as well as how these fires can be adequately prevented and protected at all times.

CAUSES OF FIRES FROM LOOSED BATTERY TERMINAL, CABLE AND WRONG ELECTRICAL CONNECTIONS:

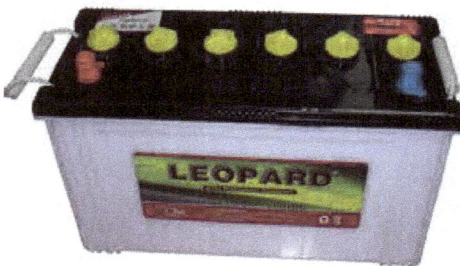

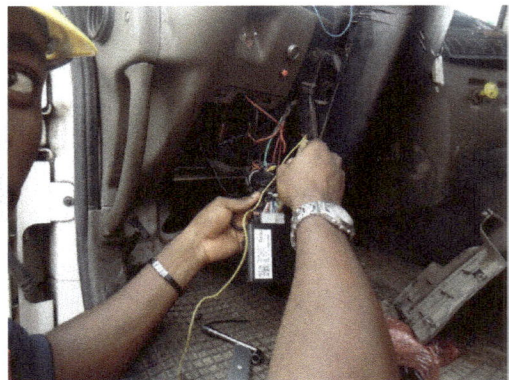

Other causes of vehicle fires have also been traced to poor handling of car batteries especially during disconnections of battery terminals and electrical connections works by repair technicians and sometimes car owners. Several electrical damages and fire have occurred through ignorant and sometimes when inadequate attention is not given during battery disconnections, removal and installations. Whenever a car battery terminals is to be loosed out or removed out of a car there are so many associated hazards attached to it, these hazards include the possibility of the positive terminal of the car battery getting in contact with the car body which is usually connected to the negative terminal of the battery [earth] any accidental contact of this parts which is usual unexpected is always deadly because of possible sparks, explosion and fire which will occur from the battery. This actions most times happens when the spanner used in loosing of the battery accidentally torches any parts of the car body. Other causes include the use of defective electrical devices and components, wrong electrical connections, sparks from loosed battery contacts e. g.   corrosion prevention, heat, loosed wires, cables and joints, when high capacity battery is used in place of low capacity one, use of low rated earth cables, using low capacity cable wires in place of recommended type this in many case will results to overheating especially during engine hard start and over loading. This problem arise due to insufficient strength of low cables and wires not been able to carry electrical loads including insufficient energy to turn engine starter motors for engine starting purpose as the case may be.

These situations when occurred will result in overheating and subsequent melting of the car cables which will eventually give rise to melting of the cable and eventually smoke will follow and bridges of the two cables will result in contact with sparks and fire will occur. Melting of cable wires can arise through overheating due to low capacity cables used in placed of recommended capacity as well as contacts between live wires and earth wires or the vehicle body which can cause short circuit [bridges]. Another important issue to note is the reverse situation in the case. If wires of higher capacity and resistance are used in replacement of low capacity types in vehicles the opposite result will occur and this would be more disastrous because of the calculated strengths and loads each cable should be able to deliver. Note that every cable wire is manufactured according to their vehicle specific loads which include the capacity of the engine, electrical appliance and components that it can carry. Cable wires of higher capacity when used for low purpose functions, instead of melting during excessive current delivery will rather allow total voltage and current through the various components and this in turn will result to burning of car electrical accessories and components, wires, circuits and possible fire outbreak. This analysis will enable us understand the importance of vehicle fuses in the prevention of damage and auto fires. The purpose of the fuse is to prevent any excessive flow of electrical current from the battery to other vehicle units, circuits and components. Excess current delivery is often caused by faulty over charging alternator and this is only prevented with the use of fuse to avoid destruction on car electrical systems and accessories, when this occurs the fuse will blow off to separate the link against the impact. Excessive flow of current can occur in vehicle due to faulty alternator components e. g when the cutout unit which helps to regulate excessive quantity of current flow from the alternator becomes bad. Over charging alternators can as well result in battery explosion, acid spill and fire and the opposite consequence could as well result to battery discharge. The above explanations should be able to provide detailed information for every responsible driver and car owners to have deep understanding about the common causes of vehicle fires as well as how they can be prevented and protected during vehicle repairs and use in other to avoid unnecessary automobile fires.

CONTROL MEASURES:

Before, during and after every vehicle repair services, vehicle owners and drivers should ensure that all necessary steps and precautions required to ensure safe vehicle maintenance are taken in to consideration before during and after car repair to avoid unnecessary car fire break e. g. proper check should be given to possible battery terminal disconnection if need be or as the case may be.

Vehicle safety and security also include how vehicle maintenance work are been carefully and safely carried out with thorough inspections on electrical and mechanical fittings at the end of every maintenance services to avoid placing cars owners and people to unexpected fire hazards.

After servicing of vehicle, it should be given thorough inspections to identify possible hidden hazards such as fuel leaks from loosed hoses or fuel lines, wrong connections, loosed battery cables and terminals, disconnections of alternator and starter motor cables, loosed bolt and nuts, loosed plug cap and high tension cables and other electrical connections as the case may be, so as to provide assurance of safety for car users after every maintenance services.

This is the fact that many vehicles have been unexpectedly burnt down because of people not been vigilant and able to maintain adequate checks or inspections after vehicle maintenance services.

The issue of soaked rags during auto repair works have also resulted in many auto fires when they are mistakenly forgotten and placed closed to hot surfaces e. g. hot exhaust pipe after vehicle maintenance works at workshop and or during vehicle fault along the high ways. These problems are usually very serious and often deadly when they occur. This challenges most often occurs after when repair works has been done to cars when people fail to have re-inspections on their car installations especially connections after repairs. Several incidents have happened where car owners would move just a few distances away from workshops and only to see their cars on smoke and ablaze. These incidents do occur when sometimes used rags are forgotten placed too closed to hot exhaust pipe or other heated parts of engine compartment. This happen when the soaked rag becomes heated in the engine compartment with increase in vapor temperature that eventually gives rise to a flash point in to actual fire.

Adequate caution should be given to all electrical joints and connections especially when coupling back disconnected and loosed wires and fittings. People should as well avoid smoking in presence of flammable substance during vehicle maintenance.

Adequate attention and caution should be given especially when battery terminals is to be loosed or disconnected, avoid spanners coming in contacts with body of vehicle especially the positive battery terminal to avoid sparks. Ensure that the earth or the negative terminal is first disconnected before the positive terminal

People should be very mindful when caring or handling flammable liquids such as petrol in presence of car batteries.

Always avoid batteries that are of higher capacity than the one in your vehicle in other to avoid unnecessary explosion and fire

Ensure that your car battery terminals have protective plastic covers or rubber insulations to avoid contact with car burnet and other metallic parts especially cars with flat burnets.

CHAPTER TEN

CAUSES OF OFFICE FIRE:

Office is an apartment, used by people or group of people, Government, public and private organizations including individuals in an organized form or structure where documentation and other official process are carried out to achieve organizations set goals.

However the use of an office apartment for the achievement of set goals by people resulted in the installation of electricity either with national grid or Power generators to enable powering and driving of both electrical and electronic equipments and other office appliances in the facilitation and execution of official task. The uses of these appliances by people, office owners, organizations, institutions, government ministries and business out fits have so far placed these ventures on different types of electrical hazards.

These hazards include;

HAZARDS AND CAUSES OF OFFICE FIRE:

>Electrocution or shock

>Electrical fire

>Ergonomic hazards included Trip, Slip & Fall hazards from trailing electrical cables and wires, wet slippery floors from water and liquid substance etc.

The case of electrocution hazards can also arise when electrical installations, connections, appliances, cables and wires such as general wiring and cable structures, fittings, use of defective air conditioners, computers, refrigerators, heaters and ring boilers, laminating machines, lightening accessories, water dispenser, etc are been poorly used and wrongly connected. Some of these hazards include situations where electrical cable insulations becomes naked or exposed and unknown to users as well as when appliances are been over loaded, using of low rated electrical cables in place of high rated capacity type, use of defective electrical appliance and equipments, defective wall sockets, using of electrical and electronic systems under wet condition or rain/ thunder lightening, using appliance when hands are wet with water and handling electrical appliances without good insulations etc.

THE USE OF SMOKE DETECTOR IN FIRE PREVENTION AT HOMES AND OFFICES.

Smoke monitor/ detector is also used in the detection of smoke in avoidance of fire, it is commonly found in offices, homes, public building, conference halls and industrial office apartments and gas storage areas etc. This device is timely installed at strategic locations to avoid dead points in buildings so as to easily gain access in the passive/detection of smell/ smoke to prevent of fires.

FIRE SAFETY SIGNS:

CONTROL MEASURES AGAINST OFFICE FIRES

Make sure that all electrical appliance are switch off when leaving office or homes.

All electrical installation works should be carried out with permit

All electrical installations should be done by competent personnel

Check list should be attached to all electrical appliances

Inspection tag should be attached to all electrical appliance and equipments in use.

All electrical cables and wire connections should be properly arranged and laid in a safe conditions

Maintain good office house keeping

Warning signs should be displayed on all equipments

All electrical appliances and panels should be well color coded and defective once should be well tagged out from good once. LOTO-

Log out and Tag procedures.

All electrical appliances should be de –energized and switch off when not in use.

Keep all work environments dry, especially your hands. Water and electricity are in compatible. A little moisture on your hand can cause serious electrocution.

Make sure that repair of all power tools must be carried out only by competent and authorized personnel

Ensure that offices are installed with smoke detector devices.

CAUSES of FILLING STATION FIRES:

There are more exposed hazards of fire at petrol production facilities, storage depots and filling stations than any other aspect in petroleum consumption. This is the facts that the nature of human activities and the manners and conducts of people in unsafe acts resulting into unsafe conditions including natural and other environmental factors surrounding these facilities is extremely very high thereby inducing several causes of petroleum liquid fires.

138

HAZARDS AND CAUSES:

Some of these hazards include, the exposure of this facility to extreme heat and ignition sources, e. g exposure to extreme heat radiation of sun light and other energy sources such as naked flames e. g   smoking, hot works, welding and maintenance works, electrical maintenance works and other heat energy sources, e. g matches, lighters, use of defective electrical appliances and wrong wiring of electrical connections, wrong installation of fuel pump connections, defective fuel pump mechanisms, inadequate isolation from ignition sources within petrol stations, static energy and leakages of fuel from tankers/trucks. Bad road conditions, poor loading of fuel trucks sometimes resulting to truck falls, Leakages during transportation as well as ignitions during discharge of petroleum products at filling stations, leakages from storage tanks, extreme fuel vapors during discharge and maintenance, improper storage and poor handling of combustible materials. These hazards have become more than what is actually expected of their controls by people as provided by law and regulations in the control of filling station fires. Some of these causes have been more predominant during sales and purchase of petroleum products by consumers [motorist and other users]. The poor attitudes of people, conducts and other forms of unsafe acts and conditions especially smoking, and poor maintenance and storage as well as during transportation of these commodities have been identified as major causes of many filling station fires. All the above causes are attributed to violations which include non compliance to filling station operational guide lines, procedures, requirements, easy people's refusal to switch off or turn off their vehicle engine before refueling at filing stations, building of filling stations without adequate assessment of the business environment [building of petrol stations to residential areas]. Some of these practices are extremely very common with especially motorist and sometimes customers who claims to be familiar with filling station attendants/operators without actually understanding the risk and dangers associated with these facilities. Several filling station fires has been investigated and recorded on ignorance and common mistakes by people and caused many filing station fires, multiple fatalities and property destructions. Not switching off vehicle engine before refueling at filling stations has also induced several petrol station fires due to heat usually generated from vehicle exhaust. How it occurs is that vehicles at idling conditions constantly releases heat and hot smoke gases from the vehicle exhaust which is capable of raising the temperature of petrol to an ignition flash point because of the volatility of petrol especially within certain atmosphere. Petrol vapor is highly ignitable when exposed to certain temperature hence the need for care to be taken when we go to refuel our vehicles at filling stations. The attitudes of petrol retailers in the process of buying and selling of these products accompanied with how petrol containers are been handled especially during petrol scarcity is also another major contributing factors that usually induces filling station fires and therefore the need for care by people to prevent occurrence of fire out break at all times.

SOME PREVENTIVE MEASURES:

- >There is need for people especially motorist and petrol users to be continually enlightened and educated of associated hazards when dealing with petroleum products at filling stations.

- >Adequate precautions need to be taken by both sales attendants and buyers of these products especially during refuel scarcity periods.

- >Caution signs should be placed and kept visible at filling stations to caution the public of fire hazards e. g No smoking signs. No phone calls to avoid distractions and actual fire etc.

- >Adequate fire protection devices should be put in place at all times e. g Fire extinguishers of recommended types to be used in the event of fire outbreak.

- >Electrical connections of all fuel pumps, fittings, accessories, lightening should be adequately installed, monitored and checked for intact all the times.

- >Naked flames and other hot work activities such as welding, repair, maintenance work should not be carried

- out near or when facility is in operations. Good isolation should be made before caring out any form of hot activities.

- >Only trained and competent personnel alone should be employed for repair maintenance services at filling stations.

- >Storage facilities should be installed at their recommended procedures and specifications as required by legislation.

- Smoking should NOT be allowed at filling station or at storage facilities.

- >Also note, fuel delivery trucks should as well be assessed for fitness and safety before they are engage for fuel delivery.

- Filling stations should also be accessed to ensure that it meets >legislative requirement in consideration to the safety of people and the environment as well as exposure of filling stations to risk of heat or hot work environment. This risk should be avoided because this can induce and trigger other forms of radiation energy in to actual fire.

- >Petrol retailers should be monitored of their unmannered attitudes during purchase of petrol especially during petrol scarcity at filling stations.

INDUCED CAUSE OF FILLING STATION FIRE with use of MOBILE PHONES [DISTRACTION]:

INTRODUCTION:

Mobile phones or hand sets are electronic devices generally used for communication purposes all over the world. Communication has been made easy because of improvement in science and technology by introducing all forms of designs of handsets with complex and non complex types. The availability and affordability of these mobile handsets have provided a safe platform for people in business and communication facilitation [Communication made easy] for people yet with associated risk attached to their use. Today despites the benefits this device is offering yet there are still associated havoc and harm that it is causing to people and the environment especially when they are been wrongly used when and where they are not supposed to be used. One of these challenges include when mobile phone is poorly used at filling stations by

143

filing station attendants especially during fuel delivery services. It has also been noticed that most vehicle fires are caused due to how sometimes heavy duty drivers unconsciously engage their mobile phones while delivering petroleum products with trucks at filling stations and during refueling of vehicles. Several havocs have being done to people, these includes major fires, explosions, burns, property destruction and deaths due to how mobile phones are been used overtime by people. One major cause of fire at filling stations is smoking and coupled with the use of mobile phone especially at filling stations, this is a major cause of distractions. Distraction has been identified as a major cause of filing station fires by mobile phones users and thirdly followed by static electrical energy which is usually generated through frictional force generated between the human body and the vehicle body when in contact with fuel vapor. The main cause of fire through distractions at filing station is that when petrol is to be delivered by filing station attendants with the use fuel pump nozzle, sometimes inadequate attention at such period comes to play and eventually result to accidental fuel spill. Petrol spills which is usually more of vapor and random in nature when occurred in the presence of ignition sources such as smoking, hot exhaust pipe of vehicle will easily induce and gives rise to fire.

CONTROL MEASURES.

- General warning! Do not smoke or make phone calls around filling station area.

- Keep all flammable substance far away from ignition source around filling station area.

- Avoid smoking at filling station

- Ensure that vehicle engine is shut down before refilling at filling stations.

- Ensure that naked flames are not placed exposed to pump nozzles

- Ensure avoidance of phone calls at filling stations due to induced energy and distractions.

- Ensure that vehicle with defective electrical and mechanical status should not be placed too closed to filling stations pumps.

- Make sure over heating engine are kept far away from pump nozzle.

- Be aware that extreme temperature can induce spontaneous fire Vehicle with overcharging alternators should not be allowed closed .

CAUSES OF SPONTANEOUS FIRES:

Different factors have been traced and identified as been responsible causes of spontaneous fires. First of all let us look in to what spontaneous fire is all about and how it occurs. Spontaneous fire occurs when waste materials of different waste compositions and classifications are disposed and deposited at same dump sites without segregation and when decayed and compressed together overtime becomes hazardous they constitutes and forms mixtures of flammable gases e. g methane gas, when the temperature of these compressed gases becomes heated overtime to their flash points the vapor of the gas becomes ignited with the help of extreme heat temperature radiated from the sunlight. This occurs when no actual flame is introduced by man. However, this kind of fire is not very common in our environments but could be common in hot climatic regions where hot atmospheric temperature is highly observed with the people on poor waste management culture.

146

CHAPTER ELEVEN

THE IMPORTANCE OF FIRST AIDER IN FIRE SAFETY:

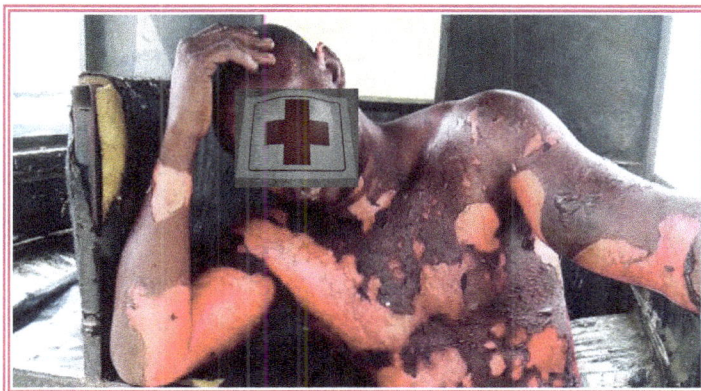

A VICTIM OF FIRE DISASTER

The essence of a first aiders in fire fighting include their roles in the intervention and response in emergency situations especially in the area of rescue, resuscitation and treatment as well as care of victims in critical life threatening situation during emergency fire outbreak. The services of first aiders in life support activities especially in cases of fire disasters have contributed immensely in rescue, resuscitation as well as facilitation in the mobilization and transportation of victim to appropriate quarters and this have saved many lives all over the globe hence the need for people to be acquainted with the basic steps in the applications of first aid so that fire victims when rescued can be promptly attended to, resuscitated and transported in the event of fire emergency situations than the all practiced ideas where people will intend to wait for real first aiders from distance place. It will go along way if resuscitation resources can be promptly made available at fire scene or situations. Lives of many fire victims have been revived and given adequate

147

support at different fire situations with the presence of real first aiders or persons with first aid knowledge.

WHO IS A FIRST AIDER?

A first aider could be referred to as someone who has being trained, certified with prove of competence and fitness to handle health and medical related emergency cases to ensure that casualty or victim conditions is adequately managed to avoid deterioration till the arrival of medical experts.

First Aid, according to the definition, it is said to be the first or initial treatment given to a victim or a casualty before the arrival of a doctor.

Emergency Response; includes all necessary steps, and processes as well as sequence of intervention in to emergency situations with regard to dully recommended procedures and guide for the rescue and resuscitation of victims in an emergency. This implies to the following which include; Assessment of incident scene to identify possible hazard, intervention procedures, rescue and resuscitation, treatment, recovery and possible evacuation of victim to hospital for adequate medical attention. However, first aid as a means of resuscitation of victims in real emergency situation have becomes importantly very useful in every aspect of human endeavors including fire emergency because of the frequently occurring fire emergency situations all over the world. However, our discussion shall be limited to those operational principles in the administration of FIRST AID and CPR in the event of actual fire out break emergency and what first aid box should look like, its roles as well as the resuscitation principles and procedures in actual fire emergency situation. Emergency as they say do not give information or signal, it is always an unplanned event which is usually accompanied with threats to life, property and environment thus, requires adequate attention and preparedness of people at all times.

Emergency situations has happened severally in the Marine, Air and road transportations with most times leaving victims without access to or assistance for rescue team, support, resuscitation and evacuation etc, e. g. response team or personnel, evacuation team, ambulance service [delay] and lack of first aid personnel to support in response, intervention, evacuation, care or treatment and facilitation. This negligence have actually resulted in several cases where fire victims would have been saved but rather victims have been kept unattended to due to sometimes non availability of first aid facilities and qualified personnel to act, coupled with ignorance and negligence, lack of knowledge in first aid and poor response procedures which in many cases have resulted to the death of many people. First aider is someone who has been trained specially on how to response and attend to victims in any emergency situations.

The first aider who is a trained personnel and always been prepared for emergency cases are sometimes been confronted with challenges of inadequate or non-availability of required materials for prompt intervention and resuscitation in actual emergency situations. To achieve effective first aid emergency response therefore requires everybody especially those of us who have actually taken time to learn about it to be always alert to ensure that prompt and quick response services are rendered at any point in time where there is emergency situations including experienced passers bye. So many loved ones have been sent to their early graves due to accident through non availability of these facilities or absents of medical attendants to provide prompt rescue to their survivals. The importance of first aid in emergency management is a very critical role in the survival of human. Another important fact to note is the competency of a First Aider!

The competence of a first aider contributes greatly to the survival of any victim in an emergency situation. A competent First Aider is an experienced person who have actually encountered several medical emergency cases, well exposed to the use of basic systems, application of basic principles, procedures and resuscitation methods that is conventionally practiced in handling of medical cases e. g application of Cardio Pulmonary Resuscitation (CPR), Chest Compression, Mouth to Mouth Breaths or Kiss Of Life, etc in the event of respiratory attack which could sometimes arise from extreme exposure of victim to actual accident or life threatening situations. These applications through first aid have saved many lives. It will be of more importance if majority of the populace are introduced to the basic principles in first aid administrations that leaving the masses in mirage even when there is real emergency scenarios that would actually require the assistance of the masses. This measure will provide a safer platform and a rescue opportunity for people in the resuscitation of victims in actual emergency scenarios. **God almighty**: human life is solely dependent on God but also in the hands of those few medical personnel or first aiders with dedicated spirit committed towards survival of humanity.

FIRST AID BOX/KIT:

Introduction:

First aid box is a medical emergency box or kit handy and mobile usually accompanied a first aider during medical emergencies. The box is usually contained with several treatment items, devices, and drugs use to administer on victim in emergency situations. The first aid box is however a box that contains several medical treatment items that are usually applied for the resuscitation and treatment of casualty during medical emergency.

ITEMS USUALLY CONTAINED IN FIRST AID BOX

The photograph above is a first aid box with some of the prescribed items that are usually contained in it.

These include the following;

Iodine
Tablets
Sterile
Needle
Plaster
Injection

Safety pin

Bandage

Antiseptic

Cotton wool
Razor blades

Handkerchief

Petroleum jelly

Adhesive tapes

Antibacterial soap

Disposable gloves

Antibiotic ointment

Aspirin/Paracetamol

Pair of scissors
Latex gloves/facial mask

- etc.

The above items are subject to expiration and should be checked frequently for expiration dates.

PREVENTIVE MEASURES:

Household members and members of the public should be trained and enlightened in the application of first aid so that at any point in time of emergency any available person and resources can be able utilized to resuscitate victims of such situations.

THE NEED OF FIRST AID BOX AT HOMES AND WORK PLACE:

First aid box should be provided in at all homes, offices and work places etc so that cases of emergency can be promptly and adequately handled in the event of fire emergencies. In modern build and constructions more especially in advance counties first aid box are basic and legal requirement for all classes of task executions. This is the reason why modern buildings and construction sites etc are always provided with first aid boxes and other live saving facilities to effectively control medical emergency cases when they eventually occur. One glaring fact is that many homes and workplaces especially in the African setting have been found without these facilities exception of some well informed house hold members. This is often because of the fact that majority of the people do not actually understand the importance and the need of these equipments at homes and work places.

MEDICAL EMERGENCY RESPONSE PROCEDURE [M E R P]

RESUSCITATING CARDIAC ARREST VICTIM IN A FIRE EMERGENCY SITUATION WITH PRACTICAL DEMONSTRATIONS OF THE PROCESS

Medical emergency is any medical condition that requires immediate medical attention. The aim is to provide help to the sick or injured person ASARP. Medical help can be rendered by a professional or trained personnel or even passer-by [First Responder]. Medical help may range from giving CPR, Stopping bleeding, immobilizing fractures, giving injection and other professional actions to identifying and raising alarm at the site or scene of an emergency, guiding on lookers from congesting the area, arranging for taxi or ambulance, and supporting other professionals [meaning everybody is an M E R team member]. This situation has call for the need for government to establish mobile emergency centers with well equipped facilities and manpower, first aiders to be located at strategic locations along major high ways and residential areas to assist people in event of any fire emergency so that victims can be easily carried to these centers for immediate

154

medical attention before referral to specific hospitals for adequate treatment than in some cases where victims of massive fire out break have been abandoned and denied rescue or response opportunity because of the absence of these facilities at strategic locations to assist in prompt intervention and resuscitation.

MEDICAL EMERGENCY RESPONSE PROCEDURES:

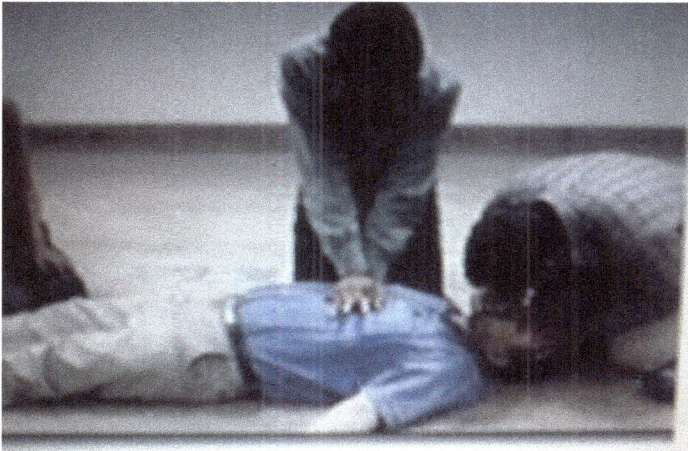

CARDIO PULMONARY RESUSCITATION [CPR]

This is a medical emergency response procedure which is given in an effort to revive a person in a cardiac arrest. It is given to unconscious and unresponsive patient or victim with no breathing or only gasps and a stopped heart.

155

METHODS/STANDARD OF APPLICATION

CPR involves chest compressions to push blood to the body tissues via the heart and blood vessels, and also to stimulate the heart to start pumping blood. It also involves artificially delivering oxygen to the victim's lungs by mouth – to – mouth breathing or mask-to- mouth ventilation with a universal chest compression and artificial rescue breath ratio of 30:2 for adult for a single person while for two persons resuscitation is a ratio of 15:2 for children and infants. **Caution;** in event of any emergency, remove all tight clothing around victim's neck. Check for fracture and bleeding and act accordingly.

BASIC STEPS AND PROCEDURE TO FOLLOW:

Briefly assessed the scene and check out for your personal safety before embarking on the response procedures.

Attend to victim with more severe case

If victim is in close hazard remove victim or remove the hazard if need be.

Check victim's responsiveness by calling his name and tapping his shoulder.

Notify emergency rescue team by calling the nurse and arrange for an ambulance to evacuate the victim.

Put victim in a comfortable position, preferably he should be lye on his back [supine position]

Maintain good air ways [by removing tight clothing and tilting the head back]

Ensure the area is well ventilated.

Kneel beside the victim to check for **CIRCULATION** and **BREATHING**.

BLOOD CIRCULATION is checked by placing your hands over the pressure points on the body preferably the carotid artery beside the neck.[Pressure points are points on the body where big arteries crosses over the bone]. If circulation is intact, you will feel the pulsation of the heart at those points.

Immediately start up Cardiac compression with a ratio of 30:2 for adults and children with one rescuer and 15:2 for children with two rescuers.

Observe if victim is breathing by feeling the exhaled air and noticing the rise and fall of the chest wall.

Start up rescue breathing mouth- to- mouth breath or mask[barrier]-to- mouth ventilation].

Check for re-establishment of breath and pulse after five [5] circles, if not present please continue with the CPR until help comes or victim is moved to health facility. But if present, put victim in a recovery position where his breathing will not be impaired.

Check for bleeding and arrest bleeding by applying pressure over the spot.

CRITICAL CONCERN TO NOTE IN EMERGENCY SITUATIONS:
In modern transportation both drivers and passengers are all entitle to have good knowledge of CPR procedures so that in event of any emergency any person can response to casualty cases to avoid total dependent on trained medical personnel from far away distance who might not be available for rescue operations in time of need. Time factor plays a very crucial role in the rescue and resuscitation of casualty therefore the need for every person to at least possess little knowledge of CPR and also to see CPR as a major tool for the resuscitation of life in an emergency.

ABOUT THE AUTHOR

The Author Safety Abadiofoni Bueseme hales from Ogu Town in the Yenagoa Local Government Area of Bayelsa State. He holds NABTEB Advance Level Mechanical Engineering, National Diploma Mechanical engineering, OND/HND in Safety [HSE] and Security Management, Trained by Daewoo Construction Nig. Ltd Ogu Base, on First Aid, CPR, Fire Fighting and Emergency Response Management and presently working with DNL 61 in the Southern Swamp Gas Gathering Pipeline Project SSGGPP as a Safety Officer HSE Department. He is the author of the books titled, [1] Industrial Safety & Emergency Prevention, [2] Causes & Prevention of Road Crashes, Safe Use of Vehicle and Maintenance all Published in the United Kingdom. The Author is happily married to Mrs Eucharia with children and leaves in Yenagoa, Bayelsa State.

159

CITATION

1. Mini hand book on fire fighting by g-platinum, accredited third party consultant to spdc/daewoo on fire prevention training.

2. Workers induction hand book on fire prevention by SPDC/DAEWOO

3. Health, safety & environmental management institute course book on fire prevention

4. On the job training with personal experience

THE END

REMEMBER THAT FIRE KILLS!

ALWAYS Avoid use of combustible materials in the presence of ignition sources AT ALL TIMES.

www.ingramcontent.com/pod-product-compliance
Lightning Source LLC
Chambersburg PA
CBHW040142200326
41519CB00032B/7585